MEDICAL DEVICE SAFETY

The Regulation of Medical Devices
for Public Health and Safety

MEDICAL DEVICE SAFETY
The Regulation of Medical Devices
for Public Health and Safety

Gordon R Higson

Institute of Physics Publishing
Bristol and Philadelphia

© IOP Publishing Ltd 2002

British Library Cataloguing-in-Publication Data

A catalogue record of this book is available from the British Library.

ISBN 0 7503 0768 4

Library of Congress Cataloging-in-Publication Data are available

Commissioning Editor: John Navas
Production Editor: Simon Laurenson
Production Control: Sarah Plenty
Cover Design: Victoria Le Billon
Marketing Executive: Laura Serratrice

Published by Institute of Physics Publishing, wholly owned by The Institute of Physics, London

Institute of Physics Publishing, Dirac House, Temple Back, Bristol BS1 6BE, UK

US Office: Institute of Physics Publishing, The Public Ledger Building, Suite 1035, 150 South Independence Mall West, Philadelphia, PA 19106, USA

Typeset by Academic + Technical, Bristol
Index by Indexing Specialists, Hove, East Sussex
Printed in the UK by Bookcraft, Midsomer Norton, Somerset

Contents

Preface

Medical devices developed during the past thirty years—such as pacemakers, hip implants, medical resonance imagers—have transformed the lives of millions of patients, often restoring their life expectancy and quality of life to that of normal healthy people.

At the same time, well-publicized problems with heart valves, breast implants and other products have given rise to unease about the safety of medical devices. The public rightly expects safety to be the first priority of both manufacturers and health authorities but has little or no appreciation that absolute safety cannot be guaranteed.

In this book I have described the approaches taken in several countries to assure the safety of medical devices. These have developed rapidly in attempts to keep pace with the astonishing rate of introduction of new devices. Although the regulations in force throughout the world appear quite different, this appearance is superficial and the fundamental approaches are very similar. I have attempted to emphasize the common and key elements to be found in the various regulatory systems and to build on them to describe a possible 'global' system.

The regulation of medical devices is both a technical and legal matter. Technical rules for safety have to be agreed by experts and to be changed as the technology develops; these rules have to be enforced by legal means. The development of international technical rules by the international standards bodies has reduced confusion and argument and is scientifically and economically beneficial.

The kind of cooperation that has been successful on the technical level is needed also in the legal field if unnecessary duplication and controversy are to be removed. Considerable success in harmonizing regulations has been achieved in Europe and South America and the harmonization process is now being actively pursued on an international level in the Global Harmonization Task Force which includes both regulatory authorities and manufacturers.

After spending some thirty years promoting international harmonization by participation in international standards committees, it is being involved in the initiation of the GHTF which gives me greatest satisfaction. The degree of commitment shown by the members has been remarkable and has led to the appearance, in an astonishingly short time, of a framework for an economical and effective world-wide system of regulation for medical devices. I am optimistic that the next decade will see such a system brought into widespread use.

Gordon R Higson
Banchory, UK
February 2001

Acknowledgments

This book is largely based on a thesis submitted for the degree of PhD at the University of Aberdeen in 2000. I wish to record my thanks to my supervisors, Professor P F Sharp and Dr A I L Campbell, for their advice and encouragement; to Quintiles–MTC for access to documents; to colleagues in Quintiles–MTC (particularly Caroline Freeman) and FDLI for the provision of documents and other information; and to my wife, Eileen, for her forbearance during the past four years.

In memoriam

Gordon Higson died suddenly on 9 August 2001 following a second stroke soon after approving the final proofs of this book.

He was born in Bolton, Lancashire in 1932 and gained a physics degree from Manchester University. After working in the aircraft and coal mining industries, he joined the Scientific and Technical Branch of the UK Department of Health and Social Security in 1969. He became Director of the Branch in 1980 and was responsible for the development of the UK approach to the regulation of medical devices.

On leaving the DHSS in 1988 he became Secretary General of the International Association of Medical Prosthesis Manufacturers (IAPM)—an association of the manufacturers of critical implantable medical devices with a particular interest in promoting an appropriate legal and regulatory environment for such devices. Here he spearheaded the IAPM's input to the development of the European Medical Devices Directives.

In 1989 he co-founded Medical Technology Consultants Europe Ltd (now part of Quintiles TransNational Corporation) and was Chairman of the company until 1995. He was a Non-executive Director of Vickers Medical from 1989 to 1993.

Gordon was a passionate advocate of national and international standards for many years. He was chairman of the BSI Healthcare Standards Committee 1983–91, Chairman of IEC Sub-Committee 62C 1982–92 and Chairman of ISO Technical Committee 210 1994–98. He was a founder of the Global Harmonization Task Force and consultant to the European Commission on Harmonization Issues 1992–96. He was also the author of *The Medical Devices Directives—a Manufacturer's Handbook* (MTC, 1st edition 1994, 2nd edition 1996) and of more than 100 papers and he made presentations at conferences world wide.

Gordon was also a family man, a constant source of support to his wife, three children and seven grandchildren. He was known for his energy, determination, humour and generosity, a man of great humanity who will be missed by family, friends and colleagues alike.

Mark Higson
Farnham Common, UK
September 2001

Foreword

In the past three decades there has been an explosion in the use of medical technologies to enhance diagnosis and improve therapy for the benefit of the patient. Associated with this expansion of the many applications of devices has been the increased attention paid by government regulatory bodies around the world to the safety and effectiveness of these products. The United States started this regulatory oversight with the 1976 Medical Device Amendments to the Federal Food, Drug and Cosmetic Act, and soon other governments, including the European Community, were developing similar programs. The various regulators' schemes have consistently been amended to accommodate the changing perspectives of safety. In addition to the formal published regulations, the various governments have made use of other guidelines and pronouncements that amplify the original regulations. Collectively these requirements may appear as a formidable hurdle for conscientious manufacturers to commercialize their products around the world.

There are numerous books and articles devoted to helping the medical device industry understand the complexities and requirements of the various regulatory bodies. The need to navigate through these international regulatory shoals is extremely important to any medical device company—but especially important to a small start up company that is attempting to 'go global'. Gordon Higson has prepared a text that clearly outlines the current regulatory situation in the European Community, the United States, Japan and other countries, and is a guide for any manufacturer planning to export into the world market.

The book describes the evolution of regulations from early treatment of devices such as pharmaceuticals to an 'engineering approach' and the use of 'key' features to assure the safety and effectiveness of the products. Higson points out that the future global system should be an evolution that incorporates the requirements of a risk-based classification scheme, application (dossier) preparation, submission, and testing and post market controls.

Quality systems, product standards, and effectiveness are discussed in separate chapters to emphasize the alternatives used in different countries. International cooperation among regulatory agencies culminated (1992) in the Global Harmonization Task Force that should set the way for the future with a universal medical device regulatory system. However, in the meantime, it is important that corporations entering into the world market understand and utilize the information that is contained in this valuable reference.

John C Villforth
President, Food and Drug Institute and former
Director, FDA's Center for Devices and Radiological Health

Chapter 1

Introduction

Medical products of all kinds have to comply with regulations to satisfy the demand for public health and safety. Medicinal products (drugs) were the first medical products to be regulated in most countries and regulations for medical devices followed—initially generally derived from drug regulations.

This book describes and examines legislation regulating the sale and use of safe medical devices. It concentrates on the situation in the European Community,[1] Japan and the USA as these three countries/regions together constitute some 85% of the world market for medical devices (HIMA 1997).[2] The regulatory situation in several other countries is briefly described. It deals with the relatively new EC Directives on medical devices in some detail because the history of their creation demonstrates that a harmonization process that many people thought would be impossible can occur successfully, and because these Directives provide the basis of a model for a future world-wide regulatory system.

Four key features of, and areas of controversy in, recent legislation are discussed at some length in Chapters 6 to 9 and their importance in the regulatory scheme is assessed. This leads to a set of proposals for a rational regulatory system which could be applied world-wide, thus simplifying the demands on manufacturers, easing technical barriers to trade, reducing the cost of meeting regulatory requirements and, by concentrating global expertise, actually increasing the level of public health and safety in so far as they are influenced by medical devices.

---

[1] The terms European Union and European Community are generally used interchangeably but, as the medical devices Directives were introduced under the EC Treaty, the terms EC and European Community will be used throughout.

[2] References are identified by a name, or initials, and a date. They will be found at the end of the thesis, listed in alphabetical and date order.

History

The history of medical device regulation is a short one and essentially begins with the passing in the United States of the Medical Device Amendments (to the Federal Food, Drug and Cosmetic Act) in 1976 (USA 76a). Before the Second World War, legislation existed in very few countries (Italy 27, USA 38), and such legislation was so general as to have little effect. In fact there was no demand for medical devices to be regulated as, until the 1950s, few medical devices existed which offered any appreciable risk to patients or users. A notable exception was X-ray equipment. The risks presented by ionizing radiation had been appreciated for some years and regulations, based on the recommendations of the International Commission on Radiological Protection (ICRP 51), governing the exposure of workers to ionizing radiation and enforcing the shielding of radiation sources, were introduced in several countries.

The next risk to be recognized was that of infection from improperly sterilized devices used for the injection or infusion of medicines or which otherwise penetrated the natural bodily defences. Legislation controlling the sale of sterilized medical devices was introduced from the 1960s, commonly under pre-existing legislation for the safety of drugs.

The only other types of medical devices receiving appreciable regulatory attention until recently were those powered by electricity, but the coverage and nature of the regulations varied from country to country.

The first really comprehensive medical device legislation in a modern form was the US Medical Device Amendments of 1976. They have had a profound influence on the design and manufacture of medical devices but, although they were updated by several revising Acts and regulations, remain somewhat limited in some ways and have been criticized by George (1994) among others. These criticisms have been addressed to some extent by the FDA Modernization Act of 1997 (USA 97a) (see Chapter 4).

The other major development in medical device regulation was the introduction of the European medical device Directives in the years 1993–98. These Directives constitute the most recent major medical device legislation in the world. The most important (in terms of its coverage) of these is the EC Directive on Medical Devices, 93/42/EEC (EC 93a), which is examined in detail in Chapter 3.

This Directive broke new ground in expanding the basic requirement that 'medical devices may be placed on the market and put into service only if they do not compromise the safety and health of patients, users and, where applicable, other persons...' into specific statements of what constitutes 'safety' in a list of 'Essential Requirements' appended to the Directive. It goes on to state that compliance with certain designated ('harmonized') standards is deemed to satisfy the legislative requirements. These are features of the 'New Approach' to product regulation in Europe

(EC 85b) and represent an engineering approach to the regulation of medical devices.

These novel features are now appearing in revised legislation being enacted in Canada, proposed in Australia, and under consideration in other countries. However, they bring the European regulations into sharp contrast with regulations, based on pharmaceutical controls, in some other countries.

The country of greatest interest is the United States which is the world's largest producer and consumer of medical devices (Wilkerson 1995) and which consequently has a major influence world-wide. The FDA Modernization Act of 1997 introduced some features which moved the US slightly towards the European approach but the major limitations still remaining are the lack of transparency in the US approval process, i.e. the criteria on which the judgements of safety and efficacy are made are not explicitly stated, and the extent to which studies of clinical effectiveness are pursued. These issues are discussed more fully later in this book.

Special features of medical products

Medical device legislation must be seen in the context of laws aimed at the protection of purchasers and users of goods of all kinds. These include laws protecting consumers against false or misleading descriptions, or imposing basic safety requirements on sellers or manufacturers of goods, such as general product safety and product liability.

General laws of this kind are supplemented by specific laws and regulations addressing certain categories of goods which are considered to present specific risks to their users, third parties or the environment. Products intended for medical purposes are specifically regulated in most, if not all, developed countries and many developing countries are currently introducing such legislation.

Medical products (both drugs and devices) are generally considered to require special measures because they are used on patients who have a lowered state of health, because they often penetrate the body's own protective barriers such as the skin, and because of their intimate connection—often over long periods—with sensitive organs and/or body fluids. Furthermore, because of their direct connection with the health of the patient, sometimes with a life-saving or life-sustaining function, it is not sufficient for a medical product simply not to cause harm (which is the case for most other products) but it must act in the way that the doctor or nurse prescribing or using it expects.

These considerations have generally led legislators to require that medical products be assessed in some way and identified as being safe and fit for use before being allowed on to the market.

Drugs versus devices

Protective legislation addressing medical products generally divides them into two distinct categories: drugs (medicinal products) and devices.

This separation is based on the different modes of action of drugs and devices which leads to quite different approaches to the establishment of their safety.

In general, drugs (medicinal products) are introduced directly into the bloodstream, the digestive system or the musculature and are conveyed to the internal organs of the body. The effects they will have on the organs, especially in repeated doses, can be predicted only roughly and may be dependent on the physical characteristics of the patient. Their true effects can only be discovered by tests on living systems: cells, animals, and ultimately human beings. These tests are aimed at establishing dose levels which are safe and effective for different types of patient, the possible presence of undesirable side effects and any contra-indications, i.e. conditions in which the drug should not be used. Side effects and contra-indications may become apparent long after the administration of the drug, and damage to organs may be irreversible. This means that patient observation must be continued for an appropriate period, also that the tests must involve appropriate numbers and types of patients in order that the information emerging from the tests should be well founded and capable of extrapolation to future patients.

By contrast, medical devices generally have only physical effects on the body; these effects are usually independent of the patient's characteristics and they stop when the use of the device is discontinued. The effects of physical actions on the body are mostly known and safe (or, at least, acceptable) levels of actions such as radiation, electric current and temperature can be checked by laboratory examination. For devices which are in intimate contact with the body, such as implants, the absence of biological effects must be proven. This will certainly involve tests on living systems but test methods are available which make it possible to avoid tests on humans in many cases.

The repeatability of the characteristics of devices and the ability to transfer them from one device to another makes it possible to codify the safety requirements for many types of device in ways which can be universally applied. These codifications are generally formulated as standards.

Standards produced by international agreement are becoming increasingly important in medical device regulation, as discussed in Chapter 7, and offer the major hope for rationalizing national and regional regulatory approaches into a world-wide system (see Chapters 7 and 10).

The existence of such standards, which allow the safety of most medical devices to be relatively easily determined in an objective manner, is a major difference between devices and drugs and marks out medical devices as engineering products.

Of course, for devices which are life-saving or life-maintaining, or which introduce new technological features, tests on humans ('clinical investigations') must be carried out. These are generally on a much smaller scale than clinical investigations on drugs and may be carried out as much to establish the function of the device as its safety.

The extent to which the regulation of medical devices should extend beyond safety—to include requirements for performance, efficacy or effectiveness—remains controversial and is one of the issues examined later in this book.

Quality systems

Any judgement of the safety and satisfaction of a medical device can only be based on the examination of an example of the device itself or of an accurate technical description of the device. For that judgement to apply to all future manufacture of that device demands that every single device must be exactly the same as the examined example or technical description.

The requirement that medical device manufacturers should implement quality systems to ensure that this is the case was pioneered by the USA in its Medical Device Amendments of 1976 and its subsequent Good Manufacturing Practice Regulation (USA 78). It was followed by the United Kingdom with its voluntary Manufacturers' Registration Scheme (Higson 1994).

The use of good manufacturing practice, or quality systems, for manufacturing has become widespread but nowadays increasing emphasis is being given to quality systems for design and manufacture based on the international standard ISO 9001 (ISO 94a) as a component of a pre-market approval system (Higson 1995). The use of quality systems in this way offers important advantages to both manufacturers and regulators and makes them one of the 'key factors' I have identified in Chapters 6, 7, 8 and 9.

Product liability

The developing regulation of medical devices has been supported by corresponding developments in product liability legislation.

In Europe the Product Liability Directive 85/374EEC (EC 85a) introduced strict liability on producers of all kinds if damage or injury is caused by a defective product. A defective product is one which 'does not provide the safety which a person is entitled to expect, taking into account all of the circumstances'. This Directive has been implemented in all Member States of the European Community. Compliance with medical device regulations may be used in defence by manufacturers.

The situation is similar in the United States, where strict liability is recognized in most States. Any violation of FDA requirements may be used as evidence against the manufacturer in US product liability litigation.

For further details of product liability law in Europe, see Hodges (1993), and for product liability in the United States, see Gleason and Speights (1994).

Globalization

The differences in the approval mechanisms from country to country are a source of inefficiency and cost. Manufacturers have long complained about the waste of effort and the delays occasioned by having to make varied submissions for approval as they enter new markets—often accompanied by repeat testing which, in the worst case, may involve different national standards or even the need to make changes in an established device.

As the industry has become more globalized in its marketing, and as more countries have introduced medical device regulations, these problems have increased and demands for a more uniform world-wide regulatory system have intensified. Despite the scepticism often expressed about the possibility of achieving a global regulatory system for medical devices, much progress towards this goal has been made in the past ten years.

The first step in this direction was made in Europe where widely disparate national regulations were scrapped and replaced by completely new regulations which apply throughout the European Community and which provide that an approval process carried out once, in any part of the EC, applies throughout the Community. This was not an easy process and took some years but, as it shows that differences can be overcome, this process is described in some detail in Chapter 2.

Another regional system has been introduced in the countries of South America as a component of the MERCOSUR trading zone and an attempt is being made to prevent the spread of disparate regulations in the Far East by the 'Asia–Pacific Harmonization Group'. However, the most significant development has been the formation of the Global Harmonization Task Force in 1992. This group has been remarkably successful so far and offers the possibility of a largely unified regulatory system within the next ten years or so. The author is optimistic about the future of this group which is discussed in detail in Chapters 10 and 11.

Definitions

The discussions in this book will be based on definitions given in the European Community Directives 65/65/EEC (medicinal products) (EC 65),

93/42/EEC (medical devices) (EC 93a) and 98/79/EC (in vitro diagnostic products) (EC 98a).

Medicinal product: Any substance or combination of substances presented for treating or preventing disease in human beings or animals. Any substance or combination of substances which may be administered to human beings or animals with a view to making a medical diagnosis or to restoring, correcting or modifying physiological functions in human beings or in animals is likewise considered a medicinal product.

Medical device: Any instrument, apparatus, appliance, material or other article, whether used alone or in combination, including the software necessary for its proper application intended by the manufacturer to be used on human beings for the purpose of:

- the monitoring, treatment or alleviation of disease,
- the diagnosis, monitoring, treatment, alleviation or compensation for an injury or handicap,
- the investigation, replacement or modification of the anatomy or of a physiological process,
- the control of conception,

and which does not achieve its principal intended action in or on the human body by pharmacological, immunological or metabolic means but which may be assisted in its function by such means.

The definition of a medical device given in EC Directive 93/42/EEC is the latest one to be found in a major piece of legislation. It drew on the definitions already existing in the Member States of the European Community and on those in other countries, particularly the USA. It is similar to other definitions of a medical device in making the prime characteristic the absence of pharmacologically-related action. A significant difference, which recognizes the growing use of drug-device combinations, is the reference to the *principal intended* action (which acknowledges that a device may be *assisted* in its function by pharmacological means).

This feature of the European definition of a medical device removes some of the difficulties of discriminating between a drug and a device, and helps with the regulation of drug–device combinations. Nevertheless, difficult cases remain and pose some of the most interesting problems in medical device legislation. Some of these issues are explored in the context of the drafting and operation of the EC Directives on medical devices in Chapters 2 and 3.

Chapter 2

The transformation in the European Community

Each of the European countries had developed its own form of medical device legislation, and these forms were very different. As an illustration, the position in the major EC countries in the early 1990s is briefly described.

United Kingdom

The United Kingdom had developed a unique form of medical device regulation based not on legislation but on the administrative provisions of the National Health Service. Health care provision outside the NHS was regarded as negligible and control of medical devices used in the NHS was seen as adequately protecting the public health.

The main instrument of advice was instructions from the Department of Health (DH) (at other times, the Department of Health and Social Security; the Ministry of Health) to Health Authorities and, in particular, the Supplies Officers of those authorities, that they should purchase only devices that complied with an appropriate British (or other comparable) Standard. Compliance with a standard was to be part of every purchasing contract and could therefore be enforced by civil contract law. Laws of general application, such as the Trade Descriptions Act 1968 (UK 68a) and the Consumer Protection Act 1987 (UK 87a), applied to such purchases, in addition to contract laws.

This system was strengthened in the 1980s by the introduction of the Manufacturers' Registration Scheme (MRS). The basis of this Scheme was that Departmental staff would inspect medical device manufacturers for compliance with a series of Guides to Good Manufacturing Practice (GMP) for different categories of device. Manufacturers who were assessed as being satisfactory were named on a Register which was issued to NHS Supplies Officers with an instruction to buy from registered manufacturers whenever possible. Registered manufacturers were regarded as being capable of making legitimate claims (in tenders or contracts) of compliance with product standards and of consistently making satisfactory devices.

The first of the Guides to Good Manufacturing Practice was published in 1981 (UK 81, Duncan 1986) and was followed by six others until by 1988 almost the entire field of medical devices was covered. As the scope of the Scheme grew and the number of manufacturers on the Register increased, it became difficult for non-registered manufacturers to sell to the National Health Service. This is believed to be the first use of GMP, or quality assurance, as a pre-market condition (the US Food and Drug Administration was already using it as a post-market control)—an approach that is discussed more fully in Chapter 6.

A handful of medical devices had been brought within the scope of the Medicines Act 1968 (UK 68b). These included products such as anaesthetic gases, absorbable sutures, collagen, hydrogels, X-ray contrast media, dental filling substances, intra-uterine contraceptive devices and contact lens fluids.

Post market controls

For approaching 40 years there has been an instruction to the National Health Service that any fault discovered in a medical device that presents, or could lead to, a hazard to a patient or to a member of staff must be reported to the Department of Health, and this instruction remains in force. Reports of defective devices are investigated and in cases where warnings are needed (about 10% of all reports) warning notices are issued on three levels of urgency varying from action required within 24 hours to advice expected to be followed. Apart from the obvious benefits of giving warning of problem devices, these reports have proved to be a fruitful source of improvements in British and international standards for medical devices—an issue that is discussed more fully in Chapters 7 and 9.

Problems with implanted devices, particularly cardiac pacemakers and heart valve prostheses, needed more systematic detection than was offered by the defect reporting procedure, and in 1978 the Department established a Pacemaker Registry and followed this with a Heart Valve Registry in 1986. These registries logged implantations and explanations of these devices and proved to be helpful in giving indications of particular types which were subject to early failure (see Chapter 9). These registries have been continued and may be extended to other types of implant.

France

The regulatory situation in France was much more complex than that in the UK.

As elsewhere, some general laws applied also to medical devices. The most important of these were the law of 21 July 1983 concerning the safety of consumers (France 83), the law of 27 December 1973 on false advertising (France 73) and the law of 31 December 1975 on the use of the French language (France 75) which required the use of the French language on all labelling and information provided on or with products.

In addition to these laws of general application there were several laws which applied only to medical devices. These laws fell into three main streams plus a number of specific product regimes.

The Code of Public Health (Code de la Santé Publique) required products within its scope to comply with the provisions of the French Pharmacopoeia. It also placed a personal responsibility on a professional pharmacist who had to be employed by the manufacturer or distributor of such products, and who was held responsible for ensuring the compliance with the Pharmacopoeia.

Although the Pharmacopoeia was primarily intended to regulate drugs, several monographs of the French Pharmacopoeia referred to medical devices. Even products not covered by specific monographs had to comply with aspects of the Pharmacopoeia, such as packaging and sterilization requirements, which applied to them. The Pharmacopoeia regime thus applied widely to medical devices.

Homologation was a specific French approval process involving technical and clinical testing of the products concerned. The procedure had been introduced by a decree of 9 December 1982 (France 82) which listed a range of medical devices which could be purchased by public hospitals only if they had been approved by the Ministry of Health. The list was extended by later decrees, but a major change was made by the law of 24 July 1987 (France 87) and its implementing decree of 1 October 1990 (France 90) which extended the requirement for the purchase of homologated products to all purchasers, thus effectively making homologation a premarket approval scheme.

The process was cumbersome and it took 6–12 months if all went well. According to Anhoury (1994) the average time for homologation was 10 months and occasioned criticism from manufacturers. If there were problems with the content of the dossiers or with the testing, the time for homologation could be much longer. It was a formidable process which was evaluated by several personnel from the Ministry of Health in a 1989 publication (Waisbord *et al.*, 1989). They had little doubt about the effectiveness of the procedure in assuring the public health.

The last major revision of the homologation procedure took place in 1993 with the issue of the *General Guide for the Homologation of Medical Products* (France 93). This Guide brought together and clarified all the current decrees and guidance into one text, as well as making some extensions. The list of devices subject to the procedure and annexed to the Guide included 70 categories of medical device.

The third strand of the French regulatory system was (and remains until the In Vitro Diagnostic Medical Device Directive is transposed into French law) a control procedure for in vitro diagnostics. France was one of the few European countries with a specific regime for such devices.

These three general schemes were supplemented by a number of regimes for the control of specific product categories. These included specific control mechanisms for contraceptive devices, medical thermometers, contact lens care products, devices susceptible of being used in an abortion, and syringes and needles. Regulations governing the installation and use of sources of ionizing radiation impacted on X-ray diagnostic and radiation therapy equipment. An even less direct control was exercised on 'heavy equipment' such as CT and MRI scanners by the law of 31 December 1970 (France 70) which required such equipment to be specially authorized as a means of controlling expenditure.

Clinical experimentation

Although not strictly a regulation of medical devices, an important measure with significant effects on the introduction of medical devices into France was the law of 20 December 1988 '*concerning the protection of patients undergoing biomedical research*' (France 88), generally known as the 'Loi Huriet' (Huriet 1998). This law imposed strict conditions on the conduct of clinical trials of all kinds, including those on medical devices. Prior authorization by a local review committee and notification to the Ministry of Health meant that the evidence of the safety of the device was thoroughly scrutinized before the trial went ahead.

Germany

The regulatory situation in Germany was even more complex than that in France. Not only were there a relatively large number of regulations but there were three Ministries and the Länder (State) authorities involved in their implementation.

The regulations described here were all developed in the former West Germany and were extended to the whole country after reunification.

As in all other countries, a series of general laws impacted on medical devices. The laws with direct application to medical devices fell into four main categories.

The German Drug Law (Arzneimittelgesetz, AMG) originated in 1976 (Germany 76a) and has been subject to several amendments. Many medical devices fell within the scope of this law by being classified by the Ministry of Youth, Family, Women and Health as either 'true' or 'fictitious' drugs.

True drugs included:

- procedure packs containing both drugs and devices
- dental filling materials
- materials of biological origin, e.g. porcine heart valves
- resorbable implants
- implants made wholly or partly of a substance, e.g. bone cement
- catheters containing a drug
- medicated dressings
- drug–device combinations of any kind.

Fictitious drugs included:

- single-use sterile instruments
- implants and assimilated products, including contact lenses, dialysers, haemofilters, plasmafilters and oxygenators
- surgical sutures and dressings
- in vitro diagnostics and disinfecting products.

The key difference between true and fictitious drugs was that true drugs had to go through a pre-market approval procedure which was generally a long and costly exercise. Other requirements on, for instance, distribution and sales and post-market surveillance were more severe for true drugs than for fictitious drugs. General requirements which applied to both true and fictitious drugs included:

- unsafe products were prohibited
- misleading information was prohibited. This referred in particular to fraudulent claims about the performance, safety or therapeutic effect of a product
- products could only be placed on the German market by an entity residing in the European Community
- components and containers of products had to comply with any relevant provisions of the German Pharmacopoeia
- products which were radioactive or which had been irradiated were prohibited unless authorized by a specific regulation. This was particularly troublesome for devices sterilized by irradiation.

The MedGV

The Regulation on the Safety of Medical Technical Equipment, referred to as the MedGV (Germany 85), dates from 14 January 1985 and came into force on 1 January 1986. This law is an extension of the Law on the Safety of Technical Equipment (Germany 68) and was administered by the Ministry of Labour and Social Affairs. It was introduced after several patient deaths from defects in radiotherapy equipment came to light and were followed by a campaign in the Press about the poor state of much of the medical equipment in German hospitals. Because of this origin, the law went well beyond the establishment of safety at the point of sale.

The scope of the regulation was defined as '*medical technical equipment, including laboratory equipment and combinations of equipment, used for medical or dental diagnostic or therapeutic purposes*', and the basic requirements were:

– safe construction of equipment
– safe installation
– instruction and training of users
– regular and careful maintenance.

The MedGV divided equipment into four groups:

Group 1   'Energized medical technical equipment as specified in the Appendix.' The Appendix listed 25 types of medical device considered to present significant risks, such as anaesthesia equipment, dialysis equipment and heart–lung machines.

Group 2   'Implantable cardiac pacemakers and other energized medical technical implants.'

Group 3   'Energized medical technical equipment not specified in the Appendix and not in Group 2.'

Group 4   'All other medical technical equipment.'

Requirements that the product should comply with the 'generally accepted technical standards', should have all controls properly marked, and should be accompanied by instructions for use, were common to all Groups. Additional specific requirements were:

– content of labelling for equipment in Groups 1, 2 and 3
– warning device for incorrect dosage for equipment in Groups 1 and 3 used for the administration of energy or drugs
– provision of an implantation record card for equipment in Group 2
– type certification for equipment in Groups 1 and 2.

The type certification requirement was a serious and difficult one. It involved two stages. The first was the provision of a sample or samples to a recognized test agency (12 were listed but they did not all have the same field of competence). A satisfactory test report then had to be submitted to the appropriate State (Land) authority which issued a type certificate (Bauartzulassung), a copy of which had to be supplied with each item of the equipment.

Paragraphs 11.3 and 15 of the MedGV placed an obligation on users to report adverse incidents with medical devices but, according to Wolf (1994), this obligation was not properly implemented in practice.

The MedGV was a complex regulation and the Ministry of Labour issued many explanatory Notices, and amended them frequently, thus making compliance with the law even more difficult.

The Regulation on the Protection from X-rays of 8 January 1987 (Germany 87) covered, among other things, X-ray installations and accelerators producing X-rays with energies between 5 keV and 3 MeV. Under this regulation X-ray tubes and housings had to be type tested for radiation leakage by the Federal Physical Science and Technology Institute. Accelerators producing particles and higher-energy X-rays were governed by the Radiation Protection Regulation of 13 October 1976 (Germany 76b). This regulation was substantially amended in 1989 to implement EC Directives on the protection of workers and members of the public from ionizing radiation.

The Verification Regulation

The Verification (Calibration) Law of 12 August 1988 (Germany 88) was administered by the Federal Ministry of the Economy. It was a wide-ranging regulation covering measuring instruments listed in 23 Appendices. Medical products were described in Appendix 15: '*Measuring instruments used in the field of health*' and Appendix 12: '*Volume measuring instruments for laboratory purposes*'.

Requirements for accuracy, stability and reliability were set down for different types of instrument, together with the approval method. Generally, at least pattern (type) approval by the Federal Physical Science and Technology Institute, possibly supported by a declaration of conformity by the manufacturer or verification by the competent authority, was required.

This discussion illustrates the particularly complex regulatory situation in Germany. Some medical devices were covered by more than one law and required approval from more than one Ministry. This made Germany probably the most difficult European country in which to market medical devices.

Italy

Italy was the first European country to legislate for medical devices as distinct from pharmaceuticals. A law on public health of 23 June 1927 (Italy 27) identified 'presidi medico-chirurgici' (medico-surgical products) as products which had to be registered with the government before they could be marketed. It stated that the products subject to this law and the rules for their registration would be published in future decrees. It was followed by further associated legislation in 1928, 1934 and 1941. All these decrees did little more than repeat the original requirement and, as none gave a definition of presidi medico-chirurgici, the legislation had little effect other than to reinforce general laws on health and safety and product liability. Serious implementation of the laws began with the publication of Circular No. 100 of 24 November 1978 (Italy 78).

The 1978 Circular identified three categories of presidi medico-chirurgici:

- for personal use (such as eye washes, contact lens solutions and insect repellents)
- for environmental use (such as insecticides and rat and mouse poisons)
- for various uses (such as resuscitation devices, intrauterine devices and orthopaedic shoes for children).

It was important for products (which extended well beyond what we now consider to be medical devices) to be placed in the right category as this determined the approval process to which they were subject. Products in the 'various use' category had to be tested in government laboratories, whereas those in other categories could be tested in private test houses. Eventually 32 products were listed as presidi medico-chirurgici but the descriptions were vague and it was generally necessary to contact the Ministry of Health to determine the regulatory status of a medical device.

For almost all medical devices, three samples had to be submitted for testing by the Higher Institute of Health (Istituto Superiore di Sanita, ISS). Their test report was considered by the Higher Council of Health which decided if the device was acceptable. Finally, a decree, signed by the Minister of Health, was published in the Official Gazette. According to Mambretti (1994) it took on average one year to obtain registration.

Electromedical devices

Although there was no legislation governing electromedical devices in Italy, the Italian Institute of the Mark of Quality (Istituto Italiano del Marchio de Qualita, IMQ) was recognized by the Ministry of Health for the testing of electromedical products and possession of the IMQ mark became important for market acceptance.

Spain

Spain had a very ambitious regulatory system for medical devices. Unfortunately the regulations were not matched by the available resources so the system never worked as intended.

The Consumer Protection Act of 1984 (Spain 84) imposed strict product liability on products including medical devices. It also contained provisions on the protection of health and safety, manufacturing, packaging, labelling and advertising.

Although it was not the first regulation covering medical devices, the basic regulation was the Royal Decree of 4 April 1978 on medical, therapeutic and corrective materials and instruments (Spain 78a) which established the basis for the classification, control and homologation of medical devices. It was followed by the General Health Law of 25 April 1986 (Spain 86), Section V of which dealt with pharmaceutical products which were deemed to include medical devices. This law enabled the authorities to exercise a wide variety of controls over medical devices, including the inspection and licensing of manufacturers and importers; the pre-market approval of products; requirements for the conduct of clinical trials and the obligation to report serious adverse events. Together with other regulations, this legislation provided for six distinct regulatory schemes:

– dressings and similar materials which had to comply with the Spanish Pharmacopoeia and, if sterile, had to be registered with the Ministry of Health;
– sterile single-use devices which required pre-market approval;
– implants. Implants were covered by the Ministerial Order of 21 July 1978 on the registration and control of clinical, therapeutic and corrective implants (Spain 78b). According to this Order, implants were to comply with technical specifications, none of which were issued. They were to be tested by, or under the supervision of, the National Centre for Pharmacobiology with respect to detailed physical, chemical and biological characteristics. These testing requirements were not appropriate for all implants and the testing facilities were not adequate. As a result, virtually no implants were approved in Spain but they were allowed into use by means of an 'exceptional import approval'. This was generally granted if a satisfactory application for pre-market approval had been made and a certificate of compliance from the country of origin was available;
– contraceptive devices which were subject to control and inspection by the Ministry of Health;
– homologation procedures which applied to a range of home-use products which could be provided under the social security system only if they had been certified as complying with requirements, and to several categories of electromedical equipment. The sale or installation in Spain of electromedical devices subject to homologation was prohibited without certification by the Ministry. This was another ambitious regulation which was not backed by sufficient resources and resort was frequently made to the exceptional approval procedure described above.
– in vitro diagnostics which, in practice, applied only to products used in the detection of HIV.

Customs control of imports

Spain was unusual in the extent to which it relied on the Customs authorities to enforce its controls on medical devices. Domestic manufacture was limited and most products were imported. Ten regulations were issued between 1983 and 1987 requiring the Customs to check that imported products had been approved or complied with the applicable laws. As the Customs rarely had the ability to check such compliance, products could be held up at the ports of entry sometimes for years and wise importers learned to make their applications for approval and to request the exceptional approval procedure.

Belgium

Non-active devices and active implants were controlled under the Royal Decree of 6 June 1960 on the Manufacture, Preparation and the Wholesale Distribution and Delivery of Medicinal Products (Belgium 60). This Decree covered surgical ligatures, sterile bandages, injection or infusion fluids together with the associated articles, and all kinds of internal prosthetic products or materials. The products had to be approved and the companies manufacturing or supplying them had to be licensed. The labelling requirements were laid down, and products had to be released on to the market by an Industrial Pharmacist registered with the Ministry of Health.

This Decree was followed by the Law on Medicines of 25 March 1964 (Belgium 64) (as amended by the laws of 16 June 1970, 21 June 1983, 22 December 1989 and 16 July 1990). This law had enabling clauses which allowed it to be applied to almost everything that would now be regarded as a medical device. In practice, these powers were rarely used. The law was enforced by the Pharmaceutical Inspectorate which had limited expertise, and did little more than continue the operation of the Decree of 1960. The two major departures from this position were in respect of ionizing radiation and in vitro diagnostics. Ionizing radiation was regulated primarily by the Royal Decree of 28 February 1963 Regulating the General Protection of the Population and Workers Against the Danger of Ionizing Radiations (Belgium 63) which required authorization of diagnostic and therapeutic radiation emitting equipment; radioisotopes used for diagnosis or therapy and medical devices sterilized by irradiation.

Netherlands

The Netherlands had a very light regulatory regime. Only two regulations were issued: one in 1970 on rubber condoms and one in 1982 on sterile medical devices.

Finland

The Ministry of Trade and Industry Resolution No. 234/84 identified electro-medical equipment as a category which had to be approved by the Electrical Inspectorate (SETI). The specific devices covered by this ruling were listed in the SETI Circular KL 118-86 (Finland 86). The list included most types of electromedical equipment, but not laboratory equipment.

Sweden

The Act of 7 May 1975 on the Control of Industrially Sterilized Single-Use Medical Devices (Sweden 75) designated the National Board of Health and Welfare (NBHW) as responsible for ensuring compliance with the Act and enabled NBHW to issue regulations for this purpose. The Board issued several directives and recommendations including a directive on labelling and recommendations on manufacture and sterilization. All sterile single-use devices had to be notified to NBHW but there was no approval process.

A series of regulations issued by the National Energy Administration required the registration of certain types of electromedical equipment. These were mainly devices for domestic use but included a few items of hospital equipment. If the compliance of the equipment with appropriate standards could not be demonstrated, testing by the Swedish Institute of Testing and Approval of Electrical Equipment (SEMKO) could be demanded.

The National Institute of Radiation Protection (SSI) issued regulations governing the safety of diagnostic X-ray equipment, dental X-ray equipment and lasers. Pre-market approval was required for all laser products.

Harmonization in Europe

It can be seen from the foregoing that by the early 1980s national regulatory systems were developing rapidly and that there was no coherence between the national approaches. The medical device industry was becoming concerned at the increasing time and expense involved in seeking separate approvals and was making known its concerns to the European Commission.[1] The preferred approach of the Commission was to press for mutual recognition of national approvals—and this approach had been enormously strengthened by the 1979 decision of the Court of Justice in the 'Cassis de Dijon'

---

[1] The European Commission is the executive and administrative arm of the European Community; the Council (of Ministers) is the decision-making body; the Court of Justice has the final word on the interpretation of Community law.

case (see page 24)—but the disparities in the national systems was too great for this approach to work and the Commission eventually came to realize that a series of harmonizing Directives would be needed.

As stated in paragraph 4 of the Commission's White Paper 'Completing the Internal Market' (EC 85c) 'The Treaty (i.e. the Treaty of Rome which established the Common Market) clearly envisaged from the outset the creation of a single integrated internal market free of restrictions on the movement of goods...'[2] but the move towards this objective was faltering by the early 1970s.

Article 28 (ex 30)[3] of the Treaties establishing the European Community and the European Union states that 'Quantitative restrictions on imports and all measures having equivalent effect shall, without prejudice to the following provisions, be prohibited between the Member States.' However, this prohibition is qualified by Article 30 (ex 36) where the measures are 'justified on grounds of public morality, public policy or public security; the protection of health and life of humans, animals or plants... Such prohibitions or restrictions shall not, however, constitute a means of arbitrary discrimination or a disguised restriction on trade between Member States.' The increasing numbers of national measures introduced under former Article 36 gave rise to suspicions that the prime aim of much of this legislation was to preserve or introduce purely national systems which would favour the home industries, rather than the protection of the public health.

Article 94 (ex 100) of the Treaties recognized the prior existence of possibly conflicting national laws and provided that 'The Council shall, acting unanimously on a proposal from the Commission, issue directives for the approximation of such provisions laid down by law, regulation or administrative action in Member States as directly affect the establishment or functioning of the common market.' The first European regulatory measure to be enacted was a Directive on 'the approximation of the rules of the Member States concerning the colouring matters authorized for use in foodstuffs intended for human consumption', which was adopted by the Council on 23 October 1962.

Recognizing that the differences between the national regulations governing medical devices were too great for a mutual recognition approach, the Community embarked upon the development of three Directives aimed at removing the conflicts between the national laws regulating the sale and marketing of medical devices.

---

[2] Article 8a of the Single European Act (EC 86) gave substance to this aspiration by stipulating that the Community should take all the necessary measures to establish the internal market by 31 December 1992.

[3] The renumbering results from amendments introduced by the Treaty of Amsterdam (EC 97a).

Directives

The Community's legislative instruments are provided by Article 249 (ex 189) of the Treaties. One of the most important of these instruments is the Directive. Its purpose is to achieve the uniformity of Community law while respecting the diversity of national legal systems. A Directive is binding on Member States with regard to the results to be achieved, but it does not define the precise method of achieving them. A Directive does not replace the laws of the Member States, but obliges them to adapt their national laws to reach the agreed Community position. The degree of freedom left to Member States is controlled by the Directive itself; some (such as those on the reduction of waste) are aimed at moving the Community gradually towards a new position and have objectives described in general terms, leaving specific provisions almost entirely to the Member States, whereas technical Directives (such as those addressed to medical devices) may be written in such detail that Member States have little room to introduce national variations (for more information on Directives and other Community instruments, see Borchardt (1994) pp. 34–41).

Furthermore, in cases where the provisions of the Directive are sufficiently clear and precise, the Directive can have direct effect if a Member State has not implemented it in national law within the period allowed or has not implemented it correctly.

The Community worked on Article 94 (ex 100) harmonizing Directives for more than 20 years without bringing many into effect. This was due to a number of causes; the principal ones, as described in paragraph 68 of the White Paper (EC 85c), were the incorporation of the technical rules in the text itself (the result being long arguments over technical details together with the possibility that by the time the Directive was agreed the rules were obsolete) and the need for unanimous agreement (the result being that any Member State could hold up a Directive by the use of its veto, which was frequently used for political purposes not connected with the Directive in question). A case in point was Directive 84/539/EEC (EC 84a).

Directive 84/539/EEC

In response to complaints from the industry supplying X-ray and other electromedical equipment about the difficulties experienced as a result of the different requirements and procedures in the Member States, the European Commission began work in 1971 on two Directives—one on radiology equipment and one on electromedical equipment.

The Radiology Equipment Directive was to have been developed around International Electrotechnical Commission (IEC) Standard 407 *Radiation protection in medical X-ray equipment 10 kV to 400 kV* (now withdrawn). It was soon realized that this standard was inadequate and, in the absence of a suitable European or international standard for the safety of X-ray equipment, the Radiology Equipment Directive was abandoned. Work was concentrated on producing a Directive to promote the free movement within the Community of electromedical equipment. This was to be an 'optional harmonization' Directive, i.e. it did not impose conditions with which all electromedical equipment had to comply, but required Member States to accept the import of equipment which complied with the specified criterion of safety. This criterion was to be compliance with the European version of the IEC general safety standard (IEC 601-1 (IEC 77)) then in an advanced state of preparation. Compliance was to be attested by the manufacturer by means of a formal declaration and the affixing of a mark (the 'reversed epsilon' mark) to the device or its packaging.

Progress within the Commission Working Group, consisting of representatives of the national regulatory authorities, the industry and the associated professions, was rapid and the formal Commission proposal (COM/74/2178) (EC 74) was issued in 1974.

It can now be seen that this was a very simple proposal by comparison with the current range of Directives. Features such as the application of one listed standard to almost the entire population of electromedical equipment, the acceptance of a manufacturer's declaration as the sole means of conformity assessment, and the absence of any post-market follow up of devices in use, would not be accepted today, but it must be remembered that, at that time, national legislation was practically non-existent and even the simple provisions of the proposal offered the regulators more security than they were accustomed to.

When the proposal was considered in the Council Working Group, consisting only of national government officials, progress initially continued to be rapid but this situation changed when reports appeared in Germany of patient deaths associated with electromedical and radiological equipment and its poor maintenance in hospitals. The German attitude hardened appreciably as a domestic response to the severe criticism appearing in the Press. The German government began work urgently on legislation which eventually appeared as the MedGV. The German representatives in the Council Working Group initially reserved their position, and then forced through a massive reduction in the scope of application of the Directive. It had been intended that the Directive would apply to all electromedical equipment except for a small number of types listed in an Annex to the Directive (a 'negative' list). Under pressure from the German delegation this was changed so that the Directive applied only to a small number of devices which were again listed, but now as a 'positive' list, in an Annex.

These changes allowed Germany to introduce the much more severe requirements of the MedGV but rendered the Directive virtually useless as a means of promoting change. However, as the adoption of the Directive needed a unanimous vote in favour, the other Member States had no option other than to agree to the German position or to abandon the Directive completely.

Although a Common Position on the emasculated Directive was reached without undue further delay, it was not voted through by the Council until several years later when a major series of negotiations released a large number of Directives held up by national vetos for one reason or another. The Directive was finally agreed on 17 September 1984 (EC 84a). It thus took some twelve years to enact a Directive of little effect—a graphic illustration of the problems the Community was then experiencing in operating a common market.

The New Approach

By the late 1970s these problems were well recognized within the Commission and the Member States and several actions were taken to address them. One of the first actions was the publication of Directive 83/189/EEC (EC 83) (repealed and replaced in 1998 (EC 98b) to extend its application to agricultural produce and medicinal products). This Directive requires Member States to notify to the Commission any draft technical regulation (other than one merely transposing the full text of an international or European standard). The Commission has the right to have the proposed regulation postponed for six months if it can show that it is likely to constitute a technical barrier to trade, and for 12 months if the Commission is preparing, or intending to prepare, a Directive on the subject. Although these delay periods are short, this has proved to be a useful Directive, and in many cases Member States have postponed new legislation indefinitely after receiving a response from the Commission. This Directive also requires the national standards bodies of the Member States to submit their standards programmes to the Commission. On the basis of this information the Commission may identify a Community need and may initiate the drafting of a European standard and delay the publication of the national standard.

However, the major actions were the Council Resolution on a 'New Approach to Technical Harmonization and Standards' (EC 85b) and the Commission White Paper on the Completion of the Internal Market (EC 85c), both published in 1985. The work was completed by the Single European Act (EC 86) signed in 1986 which committed the Community to enact the measures, described below, to establish the internal market—an area without internal frontiers within which there is free movement of goods, services, capital and persons—and set the date of 31 December 1992 for its completion.

The White Paper identified all the barriers to the operation of the European Community as a single, unified market. Differing national regulations were only one of these barriers. It proposed a programme of some 300 Directives to remove these barriers, in areas such as:

–   frontier controls (products and persons)
–   technical regulations on products
–   public procurement
–   free movement for labour and professional people
–   financial services
–   transport
–   capital movements
–   company law
–   intellectual and industrial property
–   taxation, sales tax (VAT) and excise duties
–   government subsidies

and proposed a target of the end of 1992 for their adoption.

The White Paper pointed out that this programme could be accomplished in the time envisaged only if major changes were made to the way in which Directives were drafted and adopted.

The changes in harmonizing Directives were described in the White Paper and approved by the Council Resolution of 8 May 1985 on a 'New Approach to Technical Harmonization and Standards' (EC 85b) which set out the following key principles for future Community Directives and which established a pattern for future technical regulations of all kinds:

•   Directives would not contain detailed technical provisions. They would list 'essential requirements' with which products must comply. The essential requirements would become legally binding and enforceable obligations when transposed into national laws.
•   The relevant detailed technical provisions would be contained in 'harmonized standards' adopted by the European standards organizations (CEN, CENELEC and ETSI, see Chapter 7).
•   Harmonized standards would be voluntary but a product which complied with applicable harmonized standards would be deemed to comply with the corresponding essential requirements.
•   A product which complied with the provisions of the Directive would be allowed free circulation throughout the Community.

Point VIII of the Resolution outlined the means of attestation of conformity that could be used in such Directives. These were later described in detail in the Council Resolution of 21 December 1989 on a Global Approach to Conformity Assessment (EC 89a) and the Council Decision of 13 December 1990 Concerning the Modules for the various phases of the Conformity Assessment Procedures (EC 90a) (amended on 22 July 1993 (EC 93b)).

A fundamental aspect of the New Approach derived from a 1979 ruling of the Court of Justice.[4] This celebrated case was concerned with the German regulation that a liqueur could only be lawfully sold as such in Germany if it had a minimum alcohol content. This was challenged by the French exporters of a product known as 'Cassis de Dijon'. In paragraph 14 of its judgement the Court stated 'There is therefore no valid reason why, provided that they have been lawfully produced and marketed in one of the Member States, alcoholic beverages should not be introduced into any other Member State.'

The Commission, in its Communication of October 1980 (EC 80a), expanded this judgement to the general case. It followed that, under the New Approach, the conformity assessment procedure prescribed in a Directive could be carried out in any Member State and need be carried out in only one Member State to have effect throughout the Community.

The Single European Act of 1986 (EC 86) modified the Treaty of Rome by introducing (among other changes) Article 100a (now Article 95 in the consolidated text of the Treaties) under which Directives identified as being necessary for the completion of the internal market are adopted by a weighted majority vote, rather than the unanimity required by Article 100 (now Article 94).

These two changes transformed the entire procedure for producing technical harmonization Directives and made the completion of the European single market by the end of 1992 a practicable proposition.[5]

The genuine novelty of the New Approach made it necessary for guidance to be prepared on several of the key features of Directives written in accordance with this principle. During the later 1980s and early 1990s the Commission issued a number of documents, referenced 'CERTIF', addressing aspects such as the transitional period, the status of harmonized standards, and the use of the safeguard clause. In 1994, several of these were gathered together and published as a *Guide to the implementation of Community Harmonization Directives based on the New Approach and the Global Approach* (EC 94a). This Guide was revised and reissued in 1998 (EC 98c) and revised again in 2000 (available on the website http://europa. eu.int/comm/enterprise/newapproach/legislation.htm).

The Guide is 'intended for a better understanding of directives based on New Approach and/or Global Approach, and for a more uniform application throughout different sectors and throughout the Community'. Much of the analysis in later chapters of this book is based on this document.

---

[4] Case 120/78 Rewe-Zentrale v. Bundesmonopolverwaltung für Branntwein (1979) ECR 649.
[5] For a fuller account of the completion of the internal market, see Lord Cockfield's book listed in the bibliography.

The role of industry in the medical device Directives

The original list of Directives needed to complete the internal market did not include medical devices but the industry quickly intervened with the Commission to have them added. In response to the fast-growing national legislation being introduced during the early 1980s, the medical device manufacturing and distribution industries had formed themselves into several Europe-wide trade associations. These were sector-specific and their strength generally reflected the size of the sector. The most powerful of these trade associations were (and still are):

- Coordination Committee of the Radiological and Electromedical Industries (COCIR), a federation of the national trade associations of Belgium, France, Germany, Italy, Netherlands, Spain, Sweden and the United Kingdom. Its scope was, as indicated by its title, electromedical and radiological equipment, and membership included both manufacturers and distributors. Among the manufacturers were the largest European makers of medical devices (Siemens and Philips) as well as the European arms of large American manufacturers such as Hewlett-Packard and General Electric.
- European Diagnostic Manufacturers Association (EDMA), a federation of the national trade associations of Austria, Belgium, Denmark, Finland, France, Germany, Greece, Ireland, Italy, Netherlands, Norway, Portugal, Spain, Sweden, Switzerland and the United Kingdom. Its scope was laboratory diagnostic equipment and reagents, and its membership included both manufacturers and distributors.
- European Confederation of Medical Devices Associations (EUCOMED), a federation of the national trade associations of Austria, Belgium, Denmark, Finland, France, Germany, Ireland, Italy, Netherlands, Norway, Portugal, Spain, Sweden, Switzerland and the United Kingdom. The scope of EUCOMED was less clearly defined than that of the other associations; it included a wide variety of 'commodity' devices centred around sterile, single-use devices. Membership included both distributors and manufacturers and it was by far the largest of all the trade associations.
- International Association of Prosthesis Manufacturers (IAPM). The IAPM, which was incorporated into EUCOMED in 1999, was quite different from the other European associations as it was not a federation of national bodies. It was originally formed as an association of a small number of manufacturers of implanted cardiac pacemakers and later expanded to include manufacturers of other long-term implantable devices. Its membership never exceeded twenty but the well-defined sphere of interest and the direct involvement of its members made it very influential.

Each of these associations responded in some way to the opportunity presented by the 1992 programme and the New Approach. EDMA and EUCOMED staged conferences to examine the situation after which EDMA developed a lobbying position that there were no serious technical barriers to trade affecting the diagnostics industry and there was therefore no need for a Directive in this area. This position was reversed around 1990 when the EDMA members realized that they would, in turn, face the same kinds of national legislation that the other sectors had already encountered.

EUCOMED formed a committee to examine the situation for devices within its scope and in 1988 it presented a report (EUCOMED 1988) to the Commission describing what it saw as the appropriate features of a Directive on non-active (i.e. non-powered) medical devices.

COCIR, recognizing that there was no medical device expertise or experience of any kind within the Commission, signed an agreement with the Commission in 1987 under which they undertook to fund a position of 'medical device expert' within DG III (Industry) and found a suitable person from one of its member companies to fill this post. The Commission matched this initiative by forming a medical device section consisting of this expert and a Commission *fonctionnaire* who was a lawyer. COCIR also formed a committee, which included the Commission medical device expert, which drafted a Directive on active (i.e. powered) devices and offered it to the Commission in 1988.

The IAPM, which represented manufacturers of the most sensitive, and hence most regulated, medical devices, responded to complaints from its members as early as 1982 when a visit was made to the Commission to explain the costs and delays involved in meeting highly divergent national regulations. This meeting produced no response, but a second approach to DG III in November 1982 resulted in a collaborative effort with the Commission (Cordonnier 1988).

In October 1983 a Symposium in Luxembourg on 'The Health Service Market in Europe' resulted in the setting-up of a Pacemaker Study Group with members from the Commission, the IAPM, the EWGCP (European Working Group on Cardiac Pacing, a sub-group of the European Society for Cardiology) and the authorities of the Member States. The Pacemaker Study Group met twice in 1984 under the chairmanship of a Commission official. The later meeting concluded that an EEC Directive was needed in order to ensure free movement of pacemakers, and also produced the first draft of a technical standard for pacemakers.

The Directive on Active Implantable Medical Devices

Following the publication of the Council Resolution on the New Approach, the IAPM drafted a New Approach Directive on pacemakers and presented it to the third meeting of the Pacemaker Study Group. The Commission refused to accept a draft with such a narrow scope and it was decided to widen the scope to include all active implantable (electro)medical devices.

The Commission then decided to re-constitute the Pacemaker Working Group as the 'Working Group on the Reduction of Technical Barriers to Trade in the Field of Electromedical Equipment.' The IAPM modified its draft Directive as required by the Commission and presented it to the Working Group on 28 September 1987. The Commission then took over the work and circulated its own draft for the first time in January 1988.

The Commission issued a series of working papers which were discussed at meetings attended by representatives of the Member States, the European trade associations and, occasionally, members of the medical profession. An agreed Commission text was proposed formally to the Council at the end of 1988 (EC 88). It was reviewed by the Economic and Social Committee (ECOSOC, one of the standing committees in the European administration) and by the European Parliament during 1989. It was also scrutinized by a Council Working Group, consisting of Member State officials with responsibility for the safety of medical devices, in 1989 and 1990. A Common Position, reflecting agreement between the Commission, the Parliament and the Council Working Group, was agreed early in 1990 and the Council Directive on the approximation of the laws of the Member States relating to active implantable medical devices (90/385/EEC) (EC90b) was adopted on 20 June 1990.

The content of this Directive is reviewed in Chapter 3 but it is important to realize that the discussions on this Directive settled many of the key issues relating to the regulation of medical devices in Europe. As it dealt with such a small and coherent group of products, the Commission used it as a simple model of the subsequent, more complex, Directives. A number of fundamental issues, common also to the later Directives, were settled at this stage. These included the definition of a medical device; the general form of the 'essential requirements' including the change from the original concept of 'essential safety requirements'; the use of 'performance' rather than 'effectiveness' as a requirement (discussed more fully in Chapter 8); the use of either type testing or the design control procedures of the International Organization for Standardization publication ISO 9001 (1987 version) as equivalent means of assuring satisfactory design; and the use of third parties, 'Notified Bodies,' to carry out the conformity assessment procedures. Mechanisms for allowing the use of clinical trials devices and custom-made devices without full conformity assessment were also agreed.

The use of third parties has aroused most comment outside the EC and it is perhaps surprising to recall that this was not a particularly difficult issue at the time. This is probably because several European countries, in particular France and Germany, already used third parties in their regulatory procedures and also because none of the Member States were prepared to accept the cost, delay and bureaucracy involved in building up their own regulatory authorities (which were generally very small) to the extent needed to operate the forthcoming Directive.

'Grandfathering' of devices already on the market and the concept of 'substantial equivalence' were also considered but rejected on the basis that the essential requirements would be described in adequate detail and that a transitional period between the adoption of the regulation and its enforcement would allow all devices to conform.

The most contentious issue was the equivalence of the quality and type testing routes to conformity assessment. Countries such as France and Germany, which had a long tradition of type testing and certification in both the regulated and non-regulated sectors, found it very difficult to accept that devices could be shown to be safe without such testing. On the other hand, the United Kingdom was not convinced that type testing bore any relation to the subsequent production and had pioneered quality assurance as a technique for ensuring the continued satisfaction of manufactured products of all kinds. The reconciliation of these opposing views was only achieved by allowing alternative methods of conformity assessment, as described in the next chapter.

The Directive came into force in an optional form on 1 January 1993. A transitional period, during which devices could be legally marketed either by compliance with the Directive or with the pre-existing national regulations lasted until 31 December 1994. From 1 January 1995 active implantable medical devices could be placed on the market and put into service in the European Community only if they complied with the provisions of this Directive.

The Medical Devices Directive

Commission work on the Medical Devices Directive began with the submission of the EUCOMED Report in June 1988 (EUCOMED 1988) and of the draft Directive by COCIR later in the same year. Both trade associations envisaged individual Directives serving their own constituencies, i.e. separate Directives for active and non-active devices (in the language of these Directives, 'active' means dependent on a source of power other than manual effort for its operation). Indeed, the Commission produced separate working documents and had some discussion meetings before deciding to merge the two drafts into one document.

The Commission expanded the working group which had reviewed the drafts of the Active Implantable Medical Devices Directive and entitled it the 'Working Group on the Reduction of Technical Barriers to Trade in Medical Equipment'. This Working Group was composed of representatives from all the leading European trade associations, the medical profession, the

European standards bodies CEN and CENELEC, the Member States and EFTA, and was chaired by the Commission.

This Working Group played an important role until the Commission made a proposal to the Council, at which stage it was dissolved. It examined all the draft Directives on medical devices and the industry members took full advantage of the opportunities to present comments and re-drafts. Individual industry representatives were entrusted by the Commission with specific tasks—the drafting of the classification rules being one example—and in some cases the Commission supplemented its own limited expertise by retaining them as consultants. Meetings continued for some two years before a text considered as satisfactory by the Commission was achieved. The main areas of difficulty were:

- the definition of a medical device. The aim was to give a clear demarcation from medicinal products while recognizing the existence (and growing importance) of drug–device combinations. The completely new problem of the regulation of drug–device combinations was intimately bound up with the definition;
- the essential requirements. These had to be truly comprehensive, forward-looking while respecting requirements already existing in some Member States, and sufficiently detailed to allow compliance assessments to be made in the absence of harmonized standards. The question of including efficacy/effectiveness as an essential requirement proved to be particularly contentious: some authorities—especially those which were part of drug regulatory departments—wanted it to be included, while others, supported by the industry representatives, considered that the cost and time involved in trying to establish efficacy/effectiveness ruled it out as a regulatory requirement and proposed that it should be left to the medical marketplace. The inclusion of performance (which is capable of being assessed objectively) and the assessment of side effects was regarded as a reasonable compromise. See Chapters 4 and 8 for further discussion of this question.
- the classification system. From the beginning it was accepted by all parties that the new regulations should vary in severity according to the risk presented by the device. Experience with the US system had shown the value of regulations graded in this way but had also shown the need for a more sophisticated method of classifying devices. See Chapter 4 for more details of the US system. Originally, a three-class system as in the USA was envisaged but this was changed to four classes during development of the Directive;
- the conformity assessment procedures. The major issue of the equivalence of quality systems and type testing had been addressed successfully in the AIMDD and the main problem was presented by some Member States which saw the Directive as imposing conditions less severe than their existing provisions.

The Commission made a formal submission to the Council in 1991 (EC 91a). It was considered by the ECOSOC and by the European Parliament in 1992. The Opinion of the Parliament suggested 62 amendments to the Commission proposal. The Commission accepted 36 of these amendments and incorporated them in a revised proposal of 28 July 1992 (EC 92a). A Common Position was reached at the end of the year and the Directive was adopted as 93/42/EEC on 14 June 1993 (EC93a).

The Directive on In Vitro Diagnostic Medical Devices

The introduction of the third of the Directives on medical devices was delayed by the time taken to reach agreement on the MDD, and a working paper was not issued until April 1993. A formal proposal by the Commission to the European Parliament and the Council was made in April 1995 (EC 95a). (Note that this proposal was addressed to the Parliament as well as to the Council; this was the only one of the medical device Directives to be subject to the co-decision procedure introduced under Article 251 (ex 189b) of the Treaties as modified by the Maastricht Treaty (EC 92b)).

It might have been expected that, as the third in a sequence of related Directives, the approval procedure would have been straightforward but it proved to be the most difficult. In the European Parliament the proposal was considered by the Committee for Economic, Monetary and Industrial Affairs (EMAC)—the lead Committee—and by the Committee for the Environment, Public Health and Consumer Affairs over a period of several months.

Both Committees sought to enlarge considerably the list of devices requiring Notified Body intervention, but their proposed extensions were reduced by the Parliament's voting in plenary session. The EMAC proposed extensive amendments to Articles 9 and 10, which deal with conformity assessment and national registration of medical device suppliers, but they were eventually ruled out because they would have had Member State Competent Authorities carrying out regulatory functions which under the New Approach are the responsibility of Notified Bodies.

The views of the Committees were considered and voted on by the Parliament in plenary session on 11 and 12 March 1996. After consideration of the Parliament's amendments, the Commission issued an Amended Proposal (EC 97b). It was this amended Proposal that was considered in the Council Working Party and which led to the finalized Directive.

A Common Position was adopted on 23 March 1998 and the Directive was agreed on 27 October 1998 (EC 98a). The content of the Directive and the variations from the MDD are discussed in Chapter 3.

This Directive was used also as a vehicle for introducing a number of changes into the texts of the pre-existing Medical Devices Directive. These changes are discussed in Chapter 3.

Commentary

The process of developing one regulatory system to apply across (initially) 12 separate States with completely diverse controls on medical devices was achieved surprisingly quickly. It shows the merits of having highly motivated people working within a well defined framework, and of combining the expertise of the industry with that of the regulatory authorities. It has been emphasized in this narrative that the industry played a full part in the Commission Working Group, often providing the wording of the final text and even leading panels to consider specific aspects (for instance the details of the classification system). Only when a text had progressed to a settled state, agreed by the participants in the Working Group, was further discussion restricted to government officials (with input from the European Parliament and the associated committees). Even at this stage, intensive lobbying ensured that industry's voice was not lost.

It perhaps should be pointed out here that the government officials involved in the finalizing and approval of the medical device Directives were from the Ministries of Health. Although there was an overall drive within the EC that a common market should be achieved, the problem was to develop a system that the public health officials in each Member State could accept as satisfying their responsibility for the health and safety of their citizens. They ended up with an agreed public health regulation and it does a disservice to these officials, who are now responsible for the operation of the Directives within their countries, to suggest (see Hastings (1996) and Chai (2000)) that they produced a trade measure. They operated under the one trade constraint that devices approved in one Member State had to be accepted in every other—this made it even more important that the final Directive was seen by them as satisfying their public health responsibility.

The process described in this chapter clearly worked, and in a reasonable time frame. This suggests that a similar approach could be used on a wider scale and, in fact, the Global Harmonization Task Force, discussed in a later chapter, is addressing regulatory issues in a similar way. The author's view is that the resultant Directives go a long way to providing a model for a regulatory system for use on a wider scale and therefore they will be described in some detail in the next chapter.

Chapter 3

The current situation: the EC Medical Devices Directive

The Medical Devices Directive (MDD) (93/42/EEC) (EC93a) is by far the most wide-ranging of the three medical device Directives, and represents the most comprehensive regulation of medical devices ever seen in Europe. Because of the interest it has aroused, the extent to which it is being adopted by other administrations, and the author's belief that it represents a model for a global system for medical devices, it is described here in considerable detail.

Basic structure of the Directive

There are 23 Articles and 12 Annexes in the Directive. The essence of the regulatory system is contained in the following Articles and Annexes.

Article 1   The Directive applies to medical devices and their accessories.

Article 2   Member States must ensure that medical devices can be marketed and used only if they are safe.

Article 3   A 'safe' device is one which complies with the essential requirements (ERs) contained in Annex I.

Article 5   A device which complies with harmonized standards is deemed to comply with the ERs.

Article 11   Devices must go through a procedure (as described in Annexes II–VII) to show that they conform to the ERs.

Article 9   The conformity assessment procedure depends on the Class of the device as determined by the Classification Rules in Annex IX.

Article 17   Devices which comply with the ERs and have gone through the appropriate conformity assessment procedure must bear the CE marking.

Article 4   Devices bearing the CE marking may circulate freely throughout the European Community. Special provisions (Annexes VIII and X) allow custom-made devices and devices for clinical investigation to be used without bearing the CE marking.

Article 8   This Safeguard Clause allows a Member State to act if a device is found to be unsafe.

The remaining Articles and Annexes establish the infrastructure for implementing the Directive, resolving difficulties and ambiguities, and adapting it to progress.

New concepts

This Directive introduced several concepts which were new in European medical device legislation. These included:

– Classification, i.e. the division of the entire population of medical devices into four categories, subject to different conformity assessment procedures, according to rules.
– The handling of drug–device combinations.
– The introduction of device performance into the criteria for acceptability (ERs).
– The requirement to carry out a risk assessment for each device.
– The need to provide clinical data relating to safety and performance for certain devices.
– The obligation for manufacturers to report adverse incidents and to monitor experience of the device in use.

These new concepts and other key features of the Directive will be examined below.

Recitals

The Directive itself is preceded by a number of Recitals ('Whereas . . . ') which give a general rationale for, and outline of, the Directive, as required by Article 190 of the EC Treaty. There are 22 Recitals to the MDD. Most of these explain the Commission's motives in bringing forward this Directive but four give important guidance on its application.

The Fourth Recital distinguishes the safety regulations of the Member States from the ways in which they finance their healthcare provision, and makes it clear that the Directive addresses only the former.

The Seventh and Eighth Recitals give important advice on the interpretation of the ERs. First, that references to 'minimizing' or 'reducing' risks must be interpreted with regard to 'technical and economic considerations'. Secondly, that the ERs 'should be applied with discretion to take account of the technological level existing at the time of design and of technological and economic considerations compatible with a high level of protection of health and safety'. Both of these Recitals, which were introduced by the European Parliament, are aimed at avoiding an unrealistic application of the Directive by Competent Authorities and Notified Bodies. They recognize that risk cannot be entirely eliminated, and that costly redesign to achieve marginal improvements may not be in the best interests of public health.

The Sixteenth Recital points out that, in case of need, it must be possible for the authorities to contact a person, established in the European Community, with responsibility for the device.[1]

Scope of the Directive

The MDD applies to all medical devices (see the definition quoted in Chapter 1, page 6) and accessories, where an accessory is defined as 'an article which while not being a device is intended specifically by its manufacturer to be used together with a device to enable it to be used in accordance with the use of the device intended by the manufacturer of the device', unless they are covered by the Active Implantable Medical Devices Directive (AIMDD) or the In Vitro Diagnostic Medical Devices Directive (IVDMDD). Specific exclusions (Article 1.4) are:

– medicinal products covered by Directive 65/65/EEC (EC 65)
– cosmetic products covered by Directive 76/768/EEC (EC 76)
– human blood, human blood products, human plasma or blood cells of human origin and devices which incorporate at the time of placing on the market such blood products, plasma or cells[2]
– transplants or tissues or cells of human origin or products incorporating or derived from tissues or cells of human origin
– transplants or tissues or cells of animal origin, unless a device is manufactured utilizing animal tissue which is rendered non-viable, or non-viable products derived from animal tissue.

Medicinal products

Despite the care that was taken in establishing the definition of a medical device, the exclusion of medicinal products is not straightforward as it is not always clear whether a product is a device or a medicinal product. The European Commission, with the assistance of an expert group drawn from Competent Authorities and industry, issued a guidance document originally in 1993 and revised in 1998 (EC 98d) addressing the difficulties.

This guidance points out that the criteria for demarcation are the intended purpose of the device and the method by which the principal intended action is achieved. By 'intended' is meant the intention of the

---

[1] A definition of an 'authorized representative' was introduced into the Directive on In Vitro Diagnostic Medical Devices and retrospectively added to the MDD.
[2] Directive 2000/70/EC amends the MDD to include medical devices incorporating derivatives of human blood or plasma (see below).

manufacturer. This clearly gives the manufacturer of certain products some ability to decide the classification of the product as a drug or as a device by the statement of intended purpose ascribed to the product and by the choice of the **principal** intended action; however, the manufacturer's claims cannot overrule scientific knowledge. The Commission's guidance is intended to restrict the freedom of manufacturers to determine, or unduly influence, the classification of a product and to assist all parties to the regulatory process.

The guidance quotes as examples of products which should generally be considered as medical devices:

- bone cement
- dental filling materials
- materials for sealing or adhesion of tissue
- resorbable materials used in osteo-synthesis
- sutures, including absorbable sutures
- soft and hard tissue scaffolds and fillers
- intra-uterine devices
- blood bags
- systems intended to preserve and treat blood
- viscoelastic materials with intended use for mechanical/physical purposes such as protection of tissues during and after surgery and separation of tissues
- gases and liquids for ocular tamponades
- cell separators, including those incorporating antibodies for cell marking
- wound dressings, which may be in the form of liquids, gels or pastes
- haemostatic products where the haemostatic effect results from the product's physical characteristics, or is due to the surface properties of the material
- concentrates for haemodialysis
- pressure reducing valves and regulators
- irrigation solutions (including those used in the eye) intended for pure mechanical rinsing
- devices such as catheters, guidewires and stents containing or incorporating radio isotopes where the radioactive isotope as such is not released into the body.

Some of these products are clearly not medicinal products but were presumably included as examples of medical devices which were regulated under drug laws in some Member States under pre-existing legislation.

The guidance lists the following as examples of products which should be considered as medical device accessories, and hence should be regulated under the MDD:

- contact lens care products
- disinfectants specifically intended for use with medical devices

- lubricants specifically intended for use together with medical devices
- skin barrier powders and pastes or other skin care products specifically intended for use together with ostomy bags
- challenge tests specifically intended to assess the tolerance to a given medical device, or its constituents
- gases used to drive cryoprobes and surgical tools.

The following examples are given of products which should generally be considered as medicinal products:

- spermicidal preparations
- gases used in anaesthesia and inhalation therapy, including their primary containers
- topical disinfectants (antiseptics) for use on patients
- haemostatic agents where the primary mode of action is not mechanical
- zinc paste for dermatological use
- water for injections, IV fluids and plasma volume expanders
- haemofiltration substitution solutions
- in-vivo diagnostic agents, e.g. X-ray contrast media, NMR enhancing agents, fluorescent ophthalmic strips for diagnostic purposes, carrier solutions to stabilize micro-bubbles for ultrasonic imaging
- gases for in-vivo diagnostic purposes, including lung function tests
- solutions for peritoneal dialysis
- antacids
- artificial tears
- fluoride dental preparations[3]
- agents for transport, nutrition and storage of organs intended for transplantation.

The Commission guidance does not cover every case but the reasoning and examples it contains provide a good basis for determining the status of any product. Cases of difficulty or dispute can be expected to arise and will fall within the remit of the Committee on Medical Devices established under Article 7 of the Directive.

Drug–device combinations

Combinations of drugs with devices have appeared in increasing numbers during the past 20 years or so and have presented difficulties for the regulatory authorities armed with legislation which did not recognize them. The MDD is the only legislation examined in the course of this study to make

---

[3] If the dental preparation has a device mode of action and the fluoride is ancillary to this mode of action, the preparation is a medical device. Some dental preparations where the fluorine level is less than 0.15% are classed as cosmetic products and fall under Directive 76/768/EEC (EC 76).

specific provision for such combination products in Articles 1.3 and 1.4 (and in the essential requirements). Three types of drug–device combinations are recognized and examples of each are given in EC 98d:

(1) A device is intended to deliver drugs which are placed on the market separately. Examples are most hypodermic syringes, infusion pumps, and nebulizers. The device is governed by the MDD and the drugs by Directive 65/65/EEC.

(2) A drug and its delivery device are placed on the market together as a single integral product. The product **as a whole** is governed by Directive 65/65/EEC but the device must conform to the relevant essential require- ments of the MDD. Examples are pre-filled syringes, nebulizers precharged with a specific medicinal product, implants containing medicinal products with the primary purpose of releasing the medicinal product, and intrauterine contraceptives whose primary purpose is to release progestogen.

(3) A device incorporates, as an integral part, a medicinal product the action of which is ancillary to that of the device. The product **as a whole** is governed by the MDD. Examples are heparin-coated catheters, bone cements containing antibiotics, blood bags containing anticoagulants or preservatives, haemostatic devices containing collagen, condoms coated with spermicides, and intrauterine contraceptives containing copper or silver.

Many products which fall within the scope of the MDD have been regulated under preceding national drug laws and, consequently, amendment of the drug laws was (and in some cases still is) needed in most Member States. This has generally been accomplished as part of the process of transposing the Directive into national laws or by separate amending legislation. For example, in the UK the Medicines Act 1968, the Medicines (Surgical Materials) Order 1971, the Medicines (Dental Filling Substances) Order 1975 and the Medicines (Specified Articles and Substances) Order 1976 were amended by The Medical Devices (Consequential Amendments— Medicines) Regulations 1994 (UK 94a).

The approach to drug–devices combinations adopted in the MDD is about as clear as is possible and has generally worked well. The discrimina- tion between drugs and devices, however, still presents occasional problems and guidance from advisory bodies is needed. Not only drugs, but even cosmetic products, have presented problems of definition, tooth whiteners being a case in point.

Personal protective equipment

Personal protective equipment (PPE) is excluded from the scope of the MDD by Article 1.6 but the definitions are such that items such as operating theatre wear (gowns, drapes, etc.) satisfy the definitions of both PPE and medical devices. As in other cases discussed above, the decision on which Directive applies must be based on the principal intended purpose of the product. Because of the potential for confusion between these two Directives, the European Commission has issued guidance on the discrimination between them (EC 94b):

– If the product is intended to be used in a medical context with the aim of protecting the patient from infection or some other threat to health and safety, then it should be regarded as a medical device (even though it may, at the same time, offer protection to the person using it). Examples are: surgical gloves, examination gloves, face masks, corrective glasses, and surgeons' gowns and hats.

– If the product is intended mainly to protect the person using it (irrespective of its possible use in a medical context), it should be regarded as a PPE regulated under Directive 89/686/EEC (EC 89b). Examples are: protective gloves, e.g. as used in a medical laboratory, clothing for protection against ionizing radiation, eye protection devices for professional use (e.g. welders), and gum shields for boxers.

Electromagnetic compatibility

The Directive on Electromagnetic Compatibility (EMC) (EC 89c) became mandatory on 1 January 1996. Article 2.2 of this Directive states that it does not apply to products covered by specific Directives which impose appropriate EMC requirements (in the case of the MDD, ERs 9.2 and 12.5), and Article 1.7 of the Medical Device Directive states that the MDD is a specific Directive within the meaning of Article 2.2 of the EMC Directive. Therefore, medical devices which comply with the MDD are automatically in compliance with the EMC Directive.

X-ray equipment

Article 1.8 of the MDD states that 'This Directive does not affect the application of Directive 80/836/Euratom nor of Directive 84/466/Euratom.' These are two Directives (EC 80b and EC 84b) relating to the installation and safe use of equipment emitting ionizing radiation, rather than to the equipment itself.

Essential requirements and harmonized standards

Article 3 of the MDD states that 'devices must meet the essential requirements which apply to them, taking account of the intended purpose of the devices concerned'. Article 5 states that 'Member States shall presume compliance with the essential requirements in respect of devices which are in conformity with the relevant national standards adopted pursuant to the harmonized standards . . . . ' This is a restatement of one of the fundamentals of the New Approach.

Essential requirements

The essential requirements (ERs) are listed in Annex I of the Directive. They take the form of six general requirements, which apply to all devices, and eight particular requirements (with 48 sub-divisions), only some of which apply, depending on the particular characteristics of the device in question. The general requirements may be paraphrased as follows:

(1) The devices must be safe; any risks must be acceptable in relation to the benefits offered by the device.
(2) The devices must be designed in accordance with latest knowledge; risk should be (preferably) eliminated, or protected against, or (least desirably) warned against.
(3) The devices must perform in accordance with the manufacturer's specification.
(4) The safety and performance must be maintained throughout the indicated lifetime of the device.
(5) The safety and performance of the devices must not be affected by reasonable conditions of transport and storage.
(6) Any side effects must be acceptable in relation to the benefits offered.

These general requirements may be seen as summarizing all that is required of a medical device in order for it to be considered acceptable—the first expansion of the basic requirement in Article 2 of the Directive. Without further expansion, however, they pose problems for manufacturers and regulators in deciding whether or not they are satisfied. This is because of the large element of subjective judgement demanded by (1), (2) and (6), and the practical difficulty of assessing lifetime characteristics as in (4). Consequently, it is difficult to follow the New Approach exactly as these requirements cannot easily be matched by harmonized standards. This will be discussed further below.

The particular ERs may be regarded as a further expansion of Article 2. They address the following topics:

– *Chemical, physical and biological properties*, with particular emphasis on biocompatibility and toxicology. Almost all medical devices are subject to this requirement—even those which come into contact only with the surface of the body. Devices used to store products which might be administered to patients (fluids, anaesthetic gases, etc.), or to administer them, are required to be compatible with those products. A particular requirement in this group addresses drug–device combinations. For the combination to be a device, the action of the drug component must be secondary to that of the device alone, but if it is intended to have an effect on the patient, the safety, quality and effectiveness of the drug component must be verified by methods appropriate to a drug.

– *Infection and microbial contamination.* This group of requirements covers particularly the manufacturing environment and the sterilization processes applied to a sterile device and its packaging to maintain sterility. Another important requirement concerns devices which are made from, or incorporate, tissues of animal origin. The requirements address the origin of the animal tissues and the viral inactivation processes to which they have been subjected.

– *Construction and environmental properties.* These address strength and stability but also include the possible effects of outside influences such as electromagnetic interference or interactions with other medical devices being used on the patient.

– *Devices with a measuring function.* Such devices must have accuracy and stability suitable for their function.

– *Protection against radiation.* The emission of radiation must be carefully controlled and reliable indications given when radiation is being emitted. Unintended radiation must be minimized, These requirements are amplified for devices which emit ionizing radiation.

– *Requirements for devices connected to, or equipped with, an energy source.* These cover the electrical power supply, the reliability of programmable electronic systems, power supply security and alarm signals for critical patient monitoring devices.

– *Protection against electrical risks.* These address electric shock in normal and single fault conditions.

– *Protection against mechanical and thermal risks.* These requirements address strength, stability, and safety from possible injury from moving parts or high temperatures.

– *Protection against risks posed to the patient by the supply of energy or substances.* These requirements are concerned particularly with control of the rate and quantity of energy or substances supplied to patients and the provision of suitable interlocks and alarms guarding against errors.

– *Information supplied by the manufacturer.* Relatively detailed requirements cover the labelling of devices and the content of instructions for use which must be supplied with a device.

Devices are required to comply with all ERs which apply to them. The general requirements apply to all devices but only certain particular requirements.

Careful consideration of the characteristics of the device is necessary to make a determination of the applicable ERs and, possibly even more so, to establish that certain requirements do **not** apply.

Compliance with briefly-worded essential requirements can be difficult to demonstrate and the method offered by the New Approach is to show compliance with 'harmonized standards.'

Harmonized standards

A harmonized standard is defined in the Recitals to the Directive as 'a technical specification (European standard or harmonization document) adopted, on a mandate from the Commission, by either or both of the European Committee for Standardization (CEN) or the European Committee for Electrotechnical Standardization (CENELEC) in accordance with Council Directive 83/189/EEC of 28 March 1983 laying down a procedure for the provision of information in the field of technical standards and regulations (EC 83), and pursuant to the general guidelines on cooperation between the Commission and these two bodies signed on 13 November 1984.'

Article 5 of the Medical Devices Directive adds to this definition by stating that the presumption of compliance (with the applicable essential requirements) shall be made in respect of devices which conform to national standards implementing harmonized standards the references of which have been published in the Official Journal of the European Communities, and where the references of those national standards have been published by the Member States.

It follows, therefore, that it is desirable for efficient working of the Directive that there should be a sufficient body of harmonized standards. The Commission began discussions with the European standards bodies CEN and CENELEC in 1989 on the formation of a standards programme relevant to the Directive.

As the numbers of products covered by this and other directives are so great, the standards bodies and the Commission have given their attention to the development, wherever possible, of standards of broad application in order to minimize the demand for individual product standards. CEN and CENELEC have defined the following hierarchy for standards:

Level 1 Generic standards covering fundamental requirements common to all or a very wide range of products.
Level 2 Group standards which deal with requirements applicable to a group of devices.
Level 3 Product-specific standards which contain requirements particular to one product or a small family of products.

The first standardization mandates were issued by the Commission to CEN and CENELEC in 1990 (EC90c, EC90d). These were issued in respect of the Directive on Active Implantable Medical Devices but took account of the (then forthcoming) Directive on Medical Devices. These first mandates were concerned mainly with Level 1 standards relating to quality systems for medical device design and manufacture, sterilization processes, biological safety and biological compatibility, terminology, symbols and labelling, basic methodology for clinical investigations, and general safety requirements for medical electrical equipment. They also requested a Level 2 standard on active implantable medical devices.

A mandate was issued to CEN in March 1991 for the development of standards for condoms and a comprehensive mandate was given to CEN/CENELEC in 1993 for a range of standards. This mandate called for Level 1 or Level 2 standards on the following subjects:

- risk analysis relating to the design of medical devices
- sterilizers and sterilization
- biocompatibility of medical devices
- animal tissue for medical devices
- nomenclature for medical devices
- medical electrical systems
- combinations of medical electrical equipment
- electromagnetic compatibility
- device-related radiation protection
- programmable electronic systems
- flammability and fire protection
- insulation protective codes
- symbols specific to electrical features
- general requirements for equipment emitting ionizing radiation.

The mandate also identified 109 devices or groups of devices for which Level 2 or Level 3 standards were needed.

In July 1995 a mandate was issued to the European Pharmacopoeia (EP) Commission for the further development of certain EP Monographs dealing with some products which fall within the scope of the MDD but which were formerly regarded as medicinal products.[4]

The latest mandate, issued in September 1999 (EC 99a), calls for new work on Level 1 standards on clinical investigations, risk management and traceability, as well as a number of Level 2 and Level 3 standards.

Lists of harmonized standards have been published in the Official Journal in October 1994, August 1995 and November 1995. A consolidated list, which included 47 standards, was published in August 1996 (EC 96a), and has since been supplemented by publications in the Official Journal in May 1997, August 1998 and October 1999 (EC 99b, EC 99c and EC 99d). A total of 162 harmonized standards relating to the MDD have now

---

[4] Article 5 of the MDD gives these monographs the status of harmonized standards.

been published. The last publication included the first harmonized standard relating to the IVDMDD. An up-to-date list is available on the website: http://europa.eu.int/comm/enterprise/newapproach/standardization/harmstds/reflist/meddevic.html

Many of the Level 1 standards produced in response to these mandates have now been identified as harmonized, but it will evidently be several years before the mandates are fulfilled.

A discussion of the place of standards in medical device regulation, in general, is given in Chapter 7.

Classification and conformity assessment

It is necessary for there to be some confirmation that a device meets the essential requirements. This confirmation is provided by a 'conformity assessment procedure'.

As in other administrations, it was decided that, because the Medical Devices Directive covers such a wide range of products, no one conformity assessment procedure would be satisfactory and that procedures of severity varying with the risk presented by the device should be introduced. Four levels of conformity assessment were agreed and the gateway to the procedure is a classification process. The process used in the US of expert panels classifying devices according to an assessment of the risk they present was regarded as too slow and cumbersome. Instead, an expert group led by a manufacturers' representative codified the risk assessment procedure into a set of rules which allow medical devices to be divided into four classes (Classes I, IIa, IIb and III), and the conformity assessment procedure for a device is determined by the class into which it falls.

The Classification Rules

The Classification Rules are given in Annex IX of the MDD. For non-active devices the classification system is based primarily on the sensitivity of the part of the body where the device is to be used. These rules may be summarized as follows:

Rules 1–4   Non-invasive devices, i.e. devices which do not touch the patient or come into contact only with intact skin, are mostly in Class I but some devices, mainly those which come into contact with body fluids or fluids for infusion, are in Classes IIa and IIb.

Rule 5   Devices invasive with respect to body orifices are classified according to the period of use: transient, Class I; short-term, Class IIa; long-term, Class IIb (definitions are given below).

Rules 6–8   Surgically-invasive devices are generally in Class IIa if intended for transient or short-term use, and in Class IIb if intended for long-term use, but there are exceptions. If the device comes into contact with the Central Circulatory System or the Central Nervous System, it falls into Class III.

Rule 13   Devices incorporating medicinal products are always in Class III.

Rules 14–18 deal with individual types of device not covered by the more general rules.

In the case of active devices, the rules are based mainly on the diagnostic or therapeutic purpose of the device, and the associated possibility of absorption of energy by the patient.

Rule 9   Therapeutic devices administering energy are in Class IIa unless they operate in a potentially hazardous way, in which case they fall into Class IIb.

Rule 10   Diagnostic devices which supply energy are generally in Class IIa, unless the energy is supplied as ionizing radiation or they monitor vital functions in critical care, in which case they are in Class IIb.

Rule 11   Devices administering/removing medicines/body substances are generally in Class IIa but in Class IIb if operating in a potentially hazardous way.

Rule 12   All other active devices are in Class I.

A number of definitions are crucial to the use of the Classification Rules and are included in Annex IX. Only those relating to duration of use are quoted here:

Transient   Normally intended for continuous use for less than 60 minutes.
Short-term   Normally intended for continuous use for not more than 30 days.
Long-term   Normally intended for continuous use for more than 30 days.

The classification of a device is intended to be carried out by the manufacturer. The manufacturer then makes a decision about the conformity assessment procedure to be applied. For all classes other than Class I, the procedure requires the intervention of a third party (a 'Notified Body', see below). Article 9.2 provides that if there is a dispute between the manufacturer and the Notified Body about the classification of a device, the matter is to be referred to the Competent Authority to which the Notified Body is subject. The author's experience is that the vast majority of devices can be classified without ambiguity but, to assist with difficult cases, the Commission has issued guidance on the use of the classification rules. This guidance (EC 96b) is now at Revision 5 and a further revision is expected to be published in the second quarter of 2001.

Article 9.3 provides for the Classification Rules to be adapted in the light of technical progress and experience.

Conformity assessment

The Council Decision of December 1990 (EC 90a), as amended in 1993 (EC93b), established a 'menu' of conformity assessment procedures, described as 'modules', from which the Commission draftsmen were to choose to include in a particular Directive.

Six procedures were chosen for use in the Medical Devices Directive, and these are described in Annexes II to VII of the Directive. These conformity assessment procedures are applied to Classes I–III as follows:

Class I

Compliance of the device with the essential requirements (ERs) of Annex I must be shown in technical documentation compiled in accordance with Annex VII. This technical documentation must be held available for possible examination by Competent Authorities. Additionally:

(a) If the Class I device is labelled 'sterile', those parts of the manufacturing process which govern its sterility must be covered by a production quality system according to Annex V which will be audited by a Notified Body.

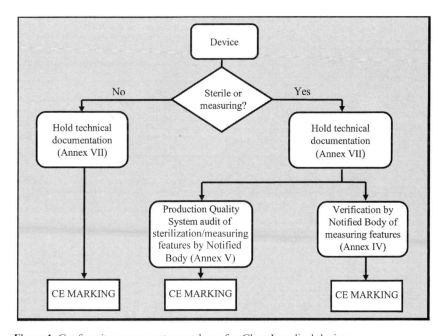

Figure 1 Conformity assessment procedures for Class I medical devices

(b) If the device has a measuring function, those aspects of the manufacturing process which govern its accuracy must be covered by a quality system to Annex IV, V or VI which will be audited by a Notified Body.

Class IIa

As for Class I devices, the design of the device, and its compliance with the ERs, must be established in technical documentation as described in Annex VII. However, for this Class, agreement of production units with the technical documentation must be assured by one of the following alternatives with the involvement of a Notified Body:

–   sample testing in accordance with Annex IV
–   a production quality system in accordance with Annex V
–   a product quality system in accordance with Annex VI.

The procedure of Annex II (quality system for design and production) is also permitted for Class IIa devices.

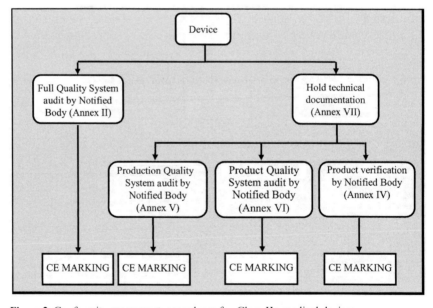

Figure 2 Conformity assessment procedures for Class IIa medical devices

Class IIb

Two alternative methods of conformity assessment are possible and the manufacturer makes the choice:

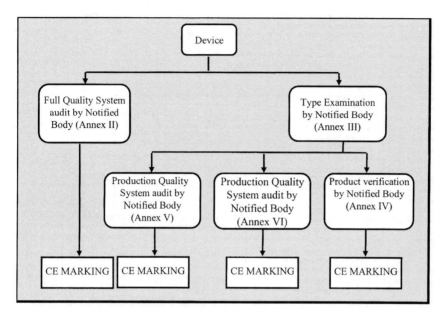

Figure 3 *Conformity assessment procedures for Class IIb medical devices*

Either   the design and manufacturing procedures must be confirmed by a Notified Body as in conformity with Annex II (quality system for design and production)

or   the design must be shown to conform to the ERs by a type examination according to Annex III carried out by a Notified Body, and the issuing of an EC type examination certificate. Compliance of production units to the approved design must be assured by a Notified Body according to Annex IV, Annex V, or Annex VI.

Class III

In the case of Class III devices, the procedures are superficially the same as for Class IIb, but significant differences are:

If Annex II is used, a design dossier, including clinical data relating to safety and performance, must be prepared for each type of device. The design dossier is examined by the Notified Body and an EC design examination certificate issued.

If Annex III is used, the documentation presented for the type examination must include clinical data relating to safety and performance. Annex VI is no longer included as a means of assuring compliance of production units.

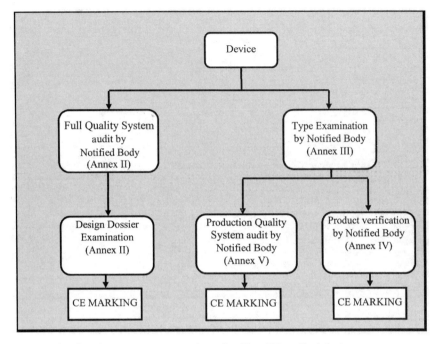

Figure 4 Conformity assessment procedures for Class III medical devices

It should be noted that all the conformity assessment Annexes, except Annex III (Type Examination) and Annex IV (EC Verification),[5] are entitled 'EC Declaration of Conformity'. They all require the manufacturer to draw up a written declaration that the products concerned meet the provisions of the Directive which apply to them. This declaration must cover a given number of identified specimens of the products manufactured and must be kept by the manufacturer.

This formulation emphasizes that the Notified Bodies can verify that the design of the device and the manufacturing conditions are satisfactory, but only the manufacturer can check all devices made and take responsibility for them. The affixing of the CE marking to every device is a further statement by the manufacturer of the conformity of the device. Section I, B(b) of the Annex to Council Decision 93/465/EEC states 'The CE marking affixed to industrial products symbolizes the fact that the natural or legal person having affixed or been responsible for the affixing of the said marking

---

[5] Annex III is an exception because it describes a process of examination of an example of a device and does not relate directly to other devices of the same type which are manufactured and placed on the market. Annex IV is an exception because it describes a procedure under which the Notified Body takes responsibility for the manufactured devices and is the only case where the Notified Body draws up a certificate of conformity.

has verified that the product conforms to all the Community total har-monization provisions which apply to it and has been the subject of the appropriate conformity evaluation procedure.'

Clinical evaluation and clinical investigation

Clinical evaluation

The final confirmation of the safety and performance of a medical device is normally provided by observation of the behaviour of the device in its intended use with patients. However, such confirmation has not generally been a feature of medical device legislation[6] and its introduction into the MDD was one of its novelties. (A more general discussion of the evaluation of clinical data in the assessment of medical devices is given in Chapter 8.)

Essential requirements 1 and 6, and in some cases 3, can only be satisfied by the evaluation of clinical data relating to the use of the device. This is pointed out in ER 14 which states 'where conformity with the essential requirements must be based on clinical data, as in Section I (6), such data must be established in accordance with Annex X' and in Annex X itself, where paragraph 1.1 reads 'As a general rule, confirmation of conformity with the requirements concerning the characteristics and performances referred to in Sections 1 and 3 of Annex I under the normal conditions of use of the device and the evaluation of the undesirable side effects must be based on clinical data in particular in the case of implantable devices and devices in Class III.'

ER 14 and Annex X are not very specific and leave some uncertainties which are causing some difficulties and inconsistencies in their application.

It is unclear just which devices require a clinical evaluation. ER 14 refers to 'Where conformity... must be based on clinical data' and Annex X states 'As a general rule...' In fact it is not generally necessary for clinical data to be evaluated in order for a decision about conformity with the essential requirements to be demonstrated. The majority of 'new' medical devices are new or updated versions of existing devices, possibly including some improved feature or incorporating some new technology in the design or the manufacture. In such cases the clinical risks or side effects presented are not changed and a technical examination is all that is required.

In the absence of guidance from the Commission, the Notified Bodies Group (see below) suggested (originally in 1996) the following cases (in addition to implantable devices and Class III devices) where clinical evidence of safety and performance can legitimately be required:

---

[6] In the United States, where the effectiveness of new devices must be proven, favourable results from clinical investigations are required, and the French homologation procedure included a period of clinical use.

- the introduction of a device using a novel technology whose features and/or mode of action are unknown
- where a device incorporates new materials coming into contact with the body or known materials being used in a new way or in a new location
- where a potentially hazardous technology is used for a new indication
- when a device approved via clinical evaluation is significantly modified.

This guidance has now been endorsed by the more formalized Notified Bodies Group and published as a Notified Bodies Recommendation (NB 98a). It is therefore likely to be generally accepted.

Annex X of the MDD allows the evaluation to be made of clinical data which may be: 'either a compilation of the relevant scientific literature currently available on the intended purpose of the device and the techniques employed as well as, if appropriate, a written report containing a critical evaluation of this compilation; or the results of all the clinical investigations made, including those carried out in conformity with Section 2', where Section 2 of the Annex describes the conditions under which clinical investigations may be carried out.

The first alternative recognizes the incremental nature of the advancement of medical devices and the continuing relevance of clinical data relating to earlier models or similar devices and Anderson (1997) has suggested that it is being accepted by Notified Bodies in the majority of cases. Guidance on the evaluation of clinical data has recently been issued by the Notified Bodies Group (NB 99).

Clinical investigations

Article 4.2 of the Medical Devices Directive allows devices intended for clinical investigation to be used, although they do not bear the CE marking, as long as they meet the conditions laid down in Article 15 and Annex VIII. These may be summarized as:

(a) Before starting an investigation, the manufacturer prepares a statement and submits it to the Competent Authorities of the countries where the investigation is to be carried out.
(b) The statement includes:

  - data to identify the device
  - an investigational plan
  - the opinion of the ethics committee concerned
  - the names of the medical specialist and institution responsible
  - a statement that the device conforms to the essential requirements apart from those aspects covered by the investigation, and that, with regard to those aspects, every precaution has been taken to protect the patients' health and safety.

(c) A technical dossier must be kept available.

(d) In the case of Class III medical devices and implantable or long-term invasive devices in Class IIa or IIb, the investigation may be started 60 days after notification if there is no intervention by the Competent Authority. Member States may stipulate a shorter period in cases where an ethics committee has agreed to the investigation.

For devices in other classes, Member States may allow the investigation to start immediately after notification if an ethics committee has agreed to the study.

(e) All investigations must be conducted in accordance with Section 2 of Annex X of the Directive. This states that the clinical investigations must be carried out in accordance with the Helsinki Declaration, as amended (Helsinki 64). Other conditions are covered much more extensively in the harmonized standard EN 540, *Clinical investigation of medical devices for human subjects* (CEN 93) (now under revision).

(f) The devices used in clinical investigations must be clearly labelled as such.

It should be noted that the provisions governing clinical investigations are a feature of the MDD where Member States are allowed some discretion in their implementation of the Directive and there is considerable variation from country to country. For example, in Denmark no application to the Competent Authority is required; in others, such as the UK, a detailed review of proposals is carried out. The provisions of the Loi Huriet continue to apply in France, and in Germany the notification must be accompanied by a statement by the sponsor that all the national requirements are met.

Notified Bodies

Article 16 of the Medical Devices Directive requires the Member States to notify the Commission and the other Member States of the bodies they designate for carrying out the conformity assessment procedures. It goes on to say: 'Member States shall apply the criteria set out in Annex XI for the designation of bodies. Bodies that meet the criteria laid down in the national standards which transpose the relevant harmonized standards shall be presumed to meet the relevant criteria.'

The criteria of Annex XI may be summarized as:

– independence from the design, manufacture, or supply of the devices in question
– integrity
– competence
– staff who are trained, experienced and able to make reports
– impartiality of staff

- possession of liability insurance
- professional secrecy.

These criteria apply whether the designated body is in the public or private sector but, in either case, the responsibility for choosing a Notified Body and ensuring that it carries out its duties properly, remains with the Competent Authority.

The 'relevant harmonized standards' amplifying these criteria are those of the EN 45000 series dealing with the accreditation and operation of certification bodies of various kinds. Seven standards in this series were adopted by the European standards bodies, CEN and CENELEC, in 1989. Those particularly relevant to the selection of Notified Bodies are EN 45001 (General criteria for the operation of testing laboratories), EN 45002 (General criteria for the assessment of testing laboratories), EN 45011 (General criteria for certification bodies operating product certification), EN 45012 (General criteria for certification bodies operating quality system certification) and EN 45004 (General criteria for the operation of various types of bodies performing inspection).

The Commission has emphasized the value of the use by Member States of the EN 45000 standards in the assessment of potential Notified Bodies in its publication CERTIF 95/2 (EC 95b), but has raised questions about whether the requirements of the standards are sufficient (EC 97c, EC 98e). These questions have been answered to some extent in Certif. 98/4 *The EN 45000 Standards, Accreditation and Notification of Notified Bodies* (EC 98f) which acknowledges weaknesses in the following areas:

- independence, impartiality and integrity
- documentation of processes
- qualification of personnel
- subcontracting
- testing activities
- coordination activities.

A work programme for addressing these weaknesses is outlined in the document. Preferably, the revision of the EN 45000 series will be carried out in conjunction with that of the corresponding ISO/IEC Guides. See also the further discussion of accreditation and certification in Chapter 6.

The Council Decision of 23 July 1993 (EC 93b) on the conformity assessment modules states: 'Notified bodies which can prove their conformity to harmonized standards (EN 45000 series) by submitting an accreditation certificate or other documentary evidence, are presumed to conform to the requirements of the directives. Member States having notified bodies unable to prove their conformity to the harmonized standards (EN 45000 series) may be requested to provide the Commission with the appropriate supporting documents on the basis of which notification was carried out.'

At the end of the 1st Quarter 2001, 62 bodies have been notified under the MDD. The national tally is:

Germany	27
UK	8
Italy	8
Norway	3
Sweden	3
Austria	2
Netherlands	2
Portugal	2
Denmark	1
Finland	1
France	1
Greece	1
Ireland	1
Luxembourg	1
Spain	1

Details of these Notified Bodies can be found on websites:

www.dimdi.de/engl/mpgengl/fr-mpgdte.htm (non-German)
www.dimdi.de/germ/mpg/bs-liste.htm (German)

This is a very large number of Notified Bodies and clearly it is difficult to ensure uniform competence and operation of such a number. The Directive does not require each Member State to have a Notified Body (or Bodies) and it is a matter of regret that it seems to have become a matter of national prestige.

It is not known whether all these bodies have been accredited against the EN 45000 standards. It is advantageous for private sector certification bodies to be able to claim such accreditation and most of them have gone through the process, but there is no such motivation for public-sector bodies and the government or quasi-government bodies which have been notified by France, Italy, Portugal, Spain and Sweden may not have gone through this procedure.

Duties of Notified Bodies

The use of private-sector bodies in the regulatory process is the most controversial feature of the European Directives. This is hardly surprising in view of the tasks to be carried out by Notified Bodies:

– Audit manufacturers' quality systems for compliance with Annexes II, V and VI.
– Examine any modifications to an approved quality system (Annexes II, V and VI).

- Carry out periodic surveillance of approved quality systems (Annexes II, V and VI).
- Examine design dossiers for Class III devices and issue EC design examination certificates (Annex II).
- Examine modifications to an approved design (Annex II).
- Carry out type examinations and issue EC type examination certificates (Annex III).
- Examine modifications to an approved type (Annex III).
- Carry out EC verification (Annex IV).
- Take measures to prevent rejected batches from reaching the market (Annex IV).
- Agree with the manufacturer time limits for the conformity assessment procedures (Article 16).
- Take into account the results of tests and verifications already carried out (Article 23).
- Communicate to other Notified Bodies (on request) all relevant information about approvals of quality systems issued, refused and withdrawn (Annexes II, V and VI).
- Communicate to other Notified Bodies (on request) all relevant information about EC type examination certificates issued, refused and withdrawn (Annex III).
- Retain information on the geographical origin of animal tissues used in medical devices (Annex I, 8.2).

The operation of Notified Bodies

The ways in which Notified Bodies are expected to work, and the constraints on their operations, are described in Chapter 6 of the 'New Approach Guide' (EC 98c).

A Notified Body must be able to carry out the entire procedure described in one of the conformity assessment Annexes. It cannot be notified for part of a procedure, but it can be notified for only a limited range of types of device. It follows that, in the case of a Class III device going through the procedure of Annex II, the examination of the design dossier must be carried out by the same Notified Body that audited the manufacturer's quality system. Conversely, where the design is assessed by type examination (Annex III), and production by one of Annexes IV, V or VI, the design and production stages may be carried out by different Notified Bodies.

Where a Notified Body is involved in the **production** stage, it must be identified by a number affixed after the CE marking. Numbers are allocated to Notified Bodies by the European Commission in the order in which they are notified, regardless of the particular Directives for which they are notified.

A Notified Body may extend the range of devices which it covers, or the range of procedures which it offers, by making use of subcontractors. The subcontractors can carry out technical tasks such as type testing or quality system audits provided that the Notified Body supplies the test or audit protocols and that the Notified Body reviews and assesses a report from the subcontractor and makes the decision on whether the requirements of the Directive are satisfied.

The use of subcontractors can also enable a Notified Body to extend its geographical coverage. Although Notified Bodies must be based within the European Economic Area, they may choose subcontractors from outside Europe as long as they can apply adequate control and supervision. Several of the larger Notified Bodies have established overseas subsidiaries as subcontractors so that the conformity assessment procedures can be carried out on the home territory of customers.

Notified Bodies are entirely responsible for the choice and the work of their subcontractors. Subcontractors must meet the same criteria of competence as the Notified Bodies themselves—this implies that they also must satisfy the applicable standards of the EN 45000 series. The Notified Bodies must continually supervise and evaluate the work of their subcontractors. They must inform the Member State which has notified them of their intention to subcontract some of their activities and keep a register of all their subcontracting.

Concerns about the uniform operation of Notified Bodies led the Commission to publish a document on coordination between Notified Bodies, Competent Authorities and the Commission in 1995 (EC 95c). This has been followed by other Commission publications (EC 97c, EC 98f) designed to address doubts about the competence of Notified Bodies, as expressed by Huriet (1996) and Kent (1996) amongst others. The Commission has also introduced a 'Code of Conduct' for Notified Bodies (EC 98g). This code would place formal obligations on both the Competent Authorities and the Notified Bodies, and would be expected to increase confidence in the Notified Body system.

A more elaborate document relating to Notified Bodies involved with medical devices (EC 00a) has been prepared by the Commission's 'Expert Group on Medical Devices'—consisting of members from the Member States' Competent Authorities, representatives from industry, and Notified Bodies themselves. It is too early yet to assess its effectiveness but this step has, in turn, been overtaken by the formation of a 'Notified Bodies Oversight Group' by the Competent Authorities. This group met for the first time in December 2000 and produced a work programme aimed at giving the Competent Authorities much greater control over the activities of the Notified Bodies.

The Notified Bodies Group

During 1992 the International Association of Prosthesis Manufacturers (IAPM) brought the Notified Bodies appointed to date together for a dialogue with industry about the application of the Directive on Active Implantable Medical Devices (AIMDD). The meetings were found to be useful and have continued with membership increasing as more Notified Bodies have been appointed, and the remit expanding to include issues relating to the MDD.

The European Organization for Testing and Certification (EOTC) has now taken over the administration of the Notified Bodies Group which is currently operating under the following terms of reference:

- to share experience and exchange views on the application of the conformity assessment procedures within the scope of the AIMDD and MDD with the aim to contribute to a uniform interpretation and application of requirements and procedures;
- to elaborate recommendations/opinions from a technical point of view related to identified matters of conformity assessment procedures;
- to give advice to the Commission following its request on identified subjects related to the application of the AIMDD and MDD;
- to consider aspects of deontology related to notified body activity and to elaborate, if necessary, statements on that topic;
- to follow harmonization activities at a European level with a close relation to the scope of the Notified Bodies Group.

The Group produces 'Consensus Statements' on issues raised by members. Statements of particular importance may be raised to the level of 'Recommendations'. These are generally referred by the Commission to its Expert Group on Medical Devices and endorsement by this Group leads to their eventual publication as 'MEDDEV' documents.

The CE Marking

Compliance with the essential requirements of the Directive and verification by a Notified Body, when required, enables the manufacturer or, in some cases, the authorized representative, to make a declaration of conformity and to mark the device with the CE marking (Article 17 of the MDD).

This use of a mark is one of the novel features of new Approach Directives. It was introduced in the Council Resolution of 1989 on the global approach to conformity assessment which states 'The affixing of the CE Mark on the products is a tangible sign of their conformity to Community rules.' Recognition of this marking affords entry to any of the Member States without further assessment for safety and, consequently, any affixing of the CE marking to a nonconforming product is prohibited by Article 18 of the MDD and is generally made an offence in the legislation transposing the Directive into national laws.

Detailed guidance on CE marking is given in Chapter 7 of the new Approach Guide (EC 98c).

Vigilance

'Vigilance' is a term introduced by the European Commission to encompass awareness of any adverse incidents involving a medical device, and a process of continually reviewing the behaviour of devices in service for signs of any potential problems (commonly known as 'post-market surveillance').

Adverse event reporting

The conformity assessment Annexes of the Medical Devices Directive, except Annex III which is concerned only with proving the design, oblige the manufacturer of all classes of device to give an undertaking to notify the Competent Authorities of incidents that involve:

'(i) any malfunction or deterioration in the characteristics and/or performance of a device, as well as any inadequacy in the labelling or the instructions for use which might lead to or might have led to the death of a patient or user or to a serious deterioration in their state of health;

(ii) any technical or medical reason connected with the characteristics or performance of a device for the reasons referred to in sub-paragraph (i) above leading to a systematic withdrawal of devices of the same type by the manufacturer.'

The Commission, in conjunction with the Member States and the leading European Trade Associations, prepared guidance on the implementation of this feature of the Directive and issued it originally in 1993 and, in revised form, in 1998 (EC 98h). This guidance was initially prepared for the Active Implantable Medical Devices Directive, which includes the same manufacturers' undertakings, but it applies equally to the MDD and the Directive on In Vitro Diagnostic Medical Devices.

This guidance interprets the reporting requirements as including:

– events which have led to death or
– life-threatening illness or injury
– permanent impairment of a body function or permanent damage to the body structure
– a condition needing intervention to prevent such permanent impairment or damage

and events which could have led to a death or serious deterioration in health but for some fortunate circumstances, or detection of a shortcoming in the device (such events are described as 'near incidents').

It goes on to say that reports should be made within the following times:

incidents   10 days
near incidents   30 days

where these are calendar days.

Reports are to be sent to the Competent Authority of the country in which the incident occurred. If the incident occurred outside the European Economic Area, but involved a device on sale or in use within the EEA, a report should be made to the Competent Authority responsible for the Notified Body involved in the assessment of the device. In the case of a Class I device (where a Notified Body is not involved), a report should be made to the Competent Authority of the Member State in which the manufacturer, or the person responsible for placing the device on the EEA market, is registered under Article 14 of the Directive.

The guidance includes a suggested format for incident reports, but this is not mandatory and several Member States have issued reporting forms. In serious cases it is preferable to report quickly by telephone or telefax and to follow up with a written report.

The guidance was revised again during 2000 to bring it into line with the recommendations of the Global Harmonization Task Force (see Chapter 10) but the new version has not yet (1st Quarter 2001) been published.

Role of the Competent Authority

Article 10 of the MDD requires Member States to establish central systems for the recording and evaluation of incidents, as described above, reported to them. It permits Member States to require medical practitioners or medical institutions to report incidents of the same kinds—and several have imposed such requirements—in which case the manufacturer, or his authorized representative established in the Community, must be informed.

The Commission guidance amplifies the role of the Competent Authority by making it clear that it is the Authority which is responsible for the resolution of any problems identified in such reports. Most commonly, the Competent Authority will confine its actions to monitoring an investigation carried out by the manufacturer. Only if the manufacturer is unable to carry out an adequate investigation will the Competent Authority investigate itself, or have an investigation carried out on its behalf.

If, after evaluation, the Competent Authority concludes that some action must be taken in respect of a reported incident, it is required by Article 10.3 to inform the Commission and the other Member States.

The 21st Recital of the MDD refers to the protection of public health being made more effective by means of 'medical device vigilance systems which are integrated at the Community level'. To fulfil this expectation, the Commission established a consortium to define the requirements for a vigilance database to be subsumed into a European regulatory data exchange system discussed in a Commission Working Paper of February 1996 (EC 96c). Article 12 of the Directive on In Vitro Diagnostic Medical Devices (IVDMDD) (see Section 4.12) took this process further by requiring Member States to establish a European databank for regulatory information. This Article was inserted into the MDD as Article 14a by Article 21.2(d) of the IVDMDD. The whole process is now reaching fruition in the EUDAMED database run on behalf of the European Commission by the German Institute for Medical Documentation and Information (DIMDI).[7]

Post-market surveillance

All the conformity assessment Annexes of the MDD (except Annex III) include a requirement that manufacturers of medical devices shall 'institute and keep up to date a systematic procedure to review experience gained from devices in the post-production phase and to implement appropriate means to apply any corrective action.'

This is a new requirement reflecting experience with some types of medical device, particularly implants, for which pre-market assessments have limitations.

Both adverse event reporting and post-market surveillance have been, or are being, adopted in many new medical device regulations. The place of these post-market controls in these regulations, and as a key feature in assuring the safety of medical devices, is discussed in Chapter 9.

The Safeguard Clause

Although Member States are required by Article 4 of the MDD to create no obstacle to the placing on the market or the putting into service of devices bearing the CE marking, they have an overriding right and duty to act to protect their populations. This power and obligation is enshrined in the Safeguard Clause (Article 8).

Under this Article a Member State is obliged to take action, which may include withdrawing the device from the market, if it establishes that a device bearing the CE marking is dangerous. A Member State which takes such action must immediately inform the Commission and this could result in extension of the action throughout the Community.

---

[7] For further information see the DIMDI website: www.dimdi.de.

The operation of the Safeguard Clause (which is found in all New Approach directives) is described in Chapter 8, Section 8.3, of the 'New Approach Guide' (EC 98c). Although there are known to have been a few instances of use of this Clause in respect of medical devices, the Commission has not released details of the numbers of incidents or of actions taken. The following analysis is therefore based on a study of the text of Article 8 of the Directive and the New Approach Guide.

Invocation of the Safeguard Clause

(a) *Discovery of a problem*
Article 2 of the Directive implies that Member States should take continuing action to check that devices being used in their territory are safe. This implication is made explicit in Chapter 8, Section 8.2, of the New Approach Guide.

While most Member States carry out market surveillance activities in the field of consumer products and services, this has not been the practice for medical devices. It is likely that Member States will generally rely on the vigilance provisions of the Directive, and reports from users, to identify problems, but some random checks are likely to be introduced—particularly on Class I products which have not been seen by a Notified Body. The UK announced in 1997 that it intended to introduce a programme of random testing of Class I devices (UK 97a) and further discussion of this programme will be found in Chapter 10, Section 10.3.2.

(b) *Establishment of a problem*
The Guide makes clear that any decision to invoke the Safeguard Clause must be based on objective and verifiable evidence that the device presents a genuine risk to the safety and health of patients, users or other persons when used for its intended purpose and that it is a systematic problem, not an isolated incident. This requirement to establish thoroughly that there is a genuine problem is supported by Article 19 of the MDD which states inter alia that any decisions to restrict the use of a device or to withdraw it from the market must state the exact grounds on which it is based. 'Such decisions shall be notified without delay to the party concerned, who shall at the same time be informed of the remedies available to him under the national law in force in the Member State in question and of the time limits to which such remedies are subject.'

(c) *Cause of a problem*
Article 8 foresees three possible causes for a CE-marked device to be found to be unsafe. These are:

– failure to meet the essential requirements
– incorrect application of the harmonized standards
– shortcomings in the standards themselves.

This cannot be viewed as an exclusive list, but it does identify systematic features of the design which would result in all devices of that design being potentially dangerous.

Application of the Safeguard Clause

The Guide makes it clear that application of the Safeguard Clause should always be regarded by a Competent Authority as a last resort. Experience of the UK voluntary reporting system has shown that most incidents are the result of non-systematic problems (including misuse of the device) and that rapid corrective action by the manufacturer is sufficient.

The Safeguard Clause should only be invoked when it is established that the CE marking on the device is not a true indication that the device is safe. This may be for one of the reasons listed in 4.8.2(c) or because the CE marking has been wrongly affixed. The case of a wrongly-affixed CE marking is discussed in Article 18 of the Directive. Where such a case is established by a Member State, the manufacturer or the authorized representative within the Community is obliged to end the infringement under conditions imposed by the Member State. If the infringement continues, the Member State must apply the Safeguard Clause.

Both Articles 18 and 19 therefore insist on the rights of the manufacturer or his representative to remedy any defect before implementation of the Safeguard Clause.

A Member State which invokes the Safeguard Clause must immediately inform the Commission. It must be able to give all necessary information to the Commission in support of its action.

The Commission must consult the 'interested parties'. These include:

– the Member State which has taken the action
– possibly, other Member States
– the manufacturer
– possibly, the Notified Body involved in the conformity assessment
– possibly, the European standards bodies
– possibly, independent experts.

From these consultations the Commission should be able to decide whether the action of the Member State was justified. If it was, then all the other Member States will be informed and the device will be removed from the market throughout the Community. There may be consequential actions for the Notified Body involved if a deficiency in the conformity assessment is found, or for the European standards bodies if the standards applied are found to have shortcomings.

If the action of the Member State is found to be not justified, the Member State and the manufacturer will be informed. The Member State will be obliged to remove its measures and to restore the device to the market. The manufacturer may be able to seek redress for consequential damage.

National transpositions

The process of transposing the MDD into national laws was completed when Belgium published its decree in March 1999 after action by the Commission in the European Court of Justice. Each country has adopted its own form of legislation, which often includes subsidiary regulations and guidance which may be of questionable status. It is important to realize that it is the national legislation that applies in each country and manufacturers wishing to sell medical devices in a particular European country must comply with the laws of that country. This is not to suggest that European harmonization has not been achieved but to bring home the fact that there are (minor) variations in details from country to country. These generally relate to the use of national languages, registration requirements for products, manufacturers and/or agents going beyond those of the Directive, procedures for obtaining permission to carry out clinical investigations and adverse event reporting.

Advice and guidance

The Medical Devices Directive provides that the Commission 'shall be assisted by' two Committees. One of these, referred to in Article 6, is a Committee set up under Directive 83/189/EEC (EC 83) on the provision of information about proposed national standards and regulations. This Committee reviews any such proposals and advises the Commission on possible actions and must be consulted by the Commission when any Safeguard Clause action is attributable to shortcomings in the standards applied, but has no special relevance to this Directive. However, Article 8 of the MDD establishes a Committee on Medical Devices which is available to examine 'any issue connected with implementation of this Directive' as well as to deal with specific issues mentioned in other Articles. These are the adaptation of the classification rules (Article 9) and derogation from the conformity assessment procedures (Article 13).

The Commission has been reluctant to set up this Committee, possibly because it consists only of representatives of the Member States, and has made use of its Expert Committee on Medical Devices, which includes representatives from manufacturers, to prepare advice on several issues. The Article 7 Committee did not meet until October 1997 and no information has been released about its activities.

Reference has been made in earlier sections of this chapter to guidance issued by the Commission on certain aspects of the MDD. Such guidance is generally prepared, at the request of the Commission, by a small group of experts and is endorsed by the Commission's Expert Group on Medical Devices before being published as a document in the 'MEDDEV' series. This Expert Group consists of representatives of the Competent Authorities of the Member States, of the major European trade associations, and of the Notified Bodies Group.

These MEDDEV documents have no legal force but, because of the degree of endorsement they have been given, they are generally accepted and widely used.

The Notified Bodies Group, besides being a source of MEDDEV documents, issues its own advice in the forms of Consensus Statements and Recommendations. These are intended for the guidance of their own members but are widely consulted by the manufacturing industry.

Most of the Member States have issued guidance on some aspect of their transposing legislation and this should be consulted by manufacturers contemplating marketing devices in specific countries. The Directive was adopted on 14 June 1993; it came into effect on an optional basis on 1 January 1995 and became mandatory on 14 June 1998. The MDD was modified by the Council Decision of 22 July 1993 (EC 93b) which introduced minor changes in the style and affixing of the CE marking, and by the Directive on In Vitro Diagnostic Medical Devices (IVDMDD) (EC 98a) which introduced some more significant changes such as:

- Article 2 now requires Member States to ensure that devices comply with the requirements of the Directive, rather than that 'they do not compromise the safety and health of patients etc.'
- Member States are allowed to set up notification systems for Class IIb and Class III devices
- a European data bank for regulatory information is established
- the transitional period for putting into service is extended to 30 June 2001 from 14 June 1998 which remains the date after which devices must bear the CE marking for placing on the market
- Notified Bodies must have sufficient expertise to assess all the types of device for which they have been notified (this reflects some of the concerns about Notified Bodies discussed earlier).

These modifications became effective from 7 June 2000.

Further modifications introduced by Directive 2000/70/EC (EC 00b) revise Article 1 of the MDD to allow the inclusion of devices which incorporate a medicinal product derived from human blood or plasma. Such devices will be controlled under the MDD in the same way as devices incorporating any other medicinal product. This amendment takes effect on 13 June 2002.

A comprehensive review of the MDD is now being undertaken by the Commission and its Experts Group. This review will address issues such as: changes in the classification rules; clarification on the need for clinical data; post-market surveillance; the Safeguard Clause and, no doubt, others, and is almost certain to result in a significant revision of the Directive.

Variations in the Directive on Active Implantable Medical Devices

The Directive on Active Implantable Medical Devices, 90/385/EEC (AIMD) (EC 90b), was the first of the three medical device directives to be finalized. It was adopted on 20 June 1990 and came into effect on 1 January 1993 with a transitional period lasting until 31 December 1994.

This Directive applies to a very small number of highly sensitive devices which were highly regulated in many EC countries. It may be viewed as a less complex version of the MDD which was used by the Commission to pioneer some of the novel features found in the later Medical Device Directive. It is a single-Class Directive in which the conformity assessment procedures correspond to those of Class III of the MDD.

Scope of the AIMDD

The Directive applies to any medical device implanted in the human body for long-term use and which depends on power, which may be either from an internal power source or supplied from outside, in order to function. It should be noted that in this Directive the definition of a medical device covers accessories used with the implantable, power-operated device and the software which may be involved in its operation. A (non-exhaustive) list of devices covered by this Directive therefore comprises:

- implantable cardiac pacemakers together with leads and electrodes
- implantable defibrillators together with leads and electrodes
- implantable nerve and muscle stimulators
- cochlear implants
- implantable drug administration devices together with catheters and sensors
- implantable physiological monitoring devices
- programmers, software and transmitters used with any of the above.

Structure of the AIMDD

The structure of the AIMDD is basically identical with that of the MDD and its essence is contained in the following Articles and Annexes:

Article 1   The Directive applies to active implantable medical devices (as defined).

Article 2   Member States must ensure that AIMDs can be marketed and used only if they are safe.

Article 3	A 'safe' device is one which complies with the essential requirements (ERs) contained in Annex 1 of the Directive.
Article 5	AIMDs which comply with harmonized standards are deemed to comply with the ERs.
Article 9	AIMDs must go through a procedure (Annexes 2–5) to show that they conform to the ERs.
Article 12	A device which complies with the ERs and has gone through the conformity assessment procedure must bear the CE marking.
Article 4	AIMDs bearing the CE marking may circulate freely throughout the EC.
Annex 1	ERs: The General Requirements include a requirement that the device must perform in accordance with its specification and that any side effects must be outweighed by the benefits offered by the device. Clinical evidence in support of these two conditions must be provided. Requirements regarding design and construction cover electrical and mechanical safety, electro-magnetic compatibility, software integrity and biocompatibility, but do not include the risk management principles found in ER 2 of the MDD (see 4.3.1).
Annexes 2–5	These Annexes describe the conformity assessment procedures available under this Directive. They are identical to those for Class III devices under the MDD and the Annexes are the same (apart from minor wording changes) as the corresponding Annexes of the MDD.

Key dates

- The Directive was adopted on 20 June 1990 and published in the Official Journal L189/17 on 20 July 1990.
- Member States were required to publish their implementing laws by 1 July 1992.
- Member States were required to bring their national laws into effect by 1 January 1993.
- For two years AIMDs could be placed on the market if they complied with either the Directive or with existing national laws.
- From 1 January 1995 CE marking of AIMDs has been mandatory.

Amendments to the Directive

The AIMDD was the first of the medical device Directives to be adopted and the later MDD and IVDMDD included provisions for amending the AIMDD to bring it into line with the two later Directives. The AIMDD was amended also by Council Directive 93/68/EEC (EC 93c) which standardizes the provisions for the affixing and use of the CE marking.

Current situation

None of the Member States succeeded in publishing its national laws by the due date of 1 July 1992, and only Germany and the United Kingdom had national legislation in place on 1 January 1993—the date when national implementing legislation was supposed to come into effect.

18 Notified Bodies have been identified for the AIMDD; these are all bodies which are also listed for the MDD.

Several of the transpositions diverge from the text of the Directive. For instance: Italy operated a transitional period only for cardiac pacemakers and requires type testing of all devices that are not made to harmonized standards; extensive notification procedures are required by Spain and UK; Portugal requires the CE marking to be accompanied by the logo of the INS.

Variations in the Directive on In Vitro Diagnostic Medical Devices

Directive 98/79/EC on In Vitro Diagnostic Medical Devices (IVDMDD) (EC98a) was agreed by the Council of Ministers on 27 October 1998. Member States were required to transpose the Directive by 7 December 1999 and bring it into effect on 7 June 2000. It will become mandatory on 7 December 2003.

Scope of the Directive

The Directive applies to in vitro diagnostic medical devices and their accessories, where an in vitro diagnostic medical device is:

'... any medical devices which is a reagent, reagent product, calibrator, control material, kit, instrument, apparatus, equipment or system, whether used alone or in combination, intended by the manufacturer to be used in vitro for the examination of specimens, including blood and tissue donations, derived from the human body, solely or principally for the purpose of providing information:
- concerning a physiological or pathological state, or
- concerning a congenital abnormality, or
- to determine the safety and compatibility with potential recipients, or
- to monitor therapeutic measures.

Specimen receptacles are considered to be in vitro diagnostic medical devices. Specimen receptacles are those devices, whether vacuum-type or not, specifically intended by their manufacturers for the primary containment and preservation of specimens derived from the human body for the purpose of in vitro diagnostic examination.

Products for general laboratory use are not in vitro diagnostic medical devices unless such products, in view of their characteristics, are specifically intended by their manufacturers to be used for in vitro diagnostic examination.'

Structure of the Directive

The structure of the Directive is similar to that of the MDD—indeed, the first eight Articles have the same titles—but there are several significant differences:

Article 9 (as Article 11 MDD) brings the classification and the conformity assessment procedures together.

Article 10 (as Article 14 MDD) makes registration mandatory for manufacturers of devices of all classes, requires the appointment of an authorized representative where a manufacturer does not have a place of business in the country concerned, and requires the submission of technical details of the devices being marketed.

Article 11 (as Article 10 MDD).

Article 12 Establishes a European data bank to hold data relating to registration, conformity assessment and vigilance reports.

Article 13 Allows Member States to impose prohibitions, restrictions or particular requirements. The Member State must then inform the Commission which will take appropriate action.

Article 14 (as Article 13 MDD).

Article 15 (as Article 16 MDD).

Article 16 (as Article 17 MDD).

Article 17 (as Article 18 MDD).

Article 18 (as Article 19 MDD).

Article 19 (as Article 20 MDD).

Article 20 Requires Competent Authorities to cooperate and exchange information.

Article 21 Introduces several amendments to the MDD, as discussed in Section 4.1. It also amends the Machinery Directive (EC 89d) to exclude all medical devices from its scope.

Essential requirements

The essential requirements follow the pattern of the MDD but with some sections restricted as there is no direct contact with patients. Requirements are added for devices for self-testing and the requirements for the information to be supplied with the product are amplified by comparison with the MDD.

Conformity assessment of in vitro diagnostic medical devices (IVDs)

The conformity assessment procedures recognize the particular nature of IVDs, i.e. that because they do not contact patients they do not have the potential to harm them directly; the danger they present is one of misdiagnosis, and only in certain cases is a misdiagnosis likely to result in a serious danger to health. Accordingly:

- For all products other than those listed in Annex II of the Directive the manufacturer follows a procedure similar to that for Class I medical devices under the MDD (holding of technical documentation) but with the significant additional requirement that the manufacturer must have a quality system for manufacture (this is not described in detail but a system to ISO 9002 would satisfy the requirement).
- If the product is intended for patient self-testing, the design must be submitted to a Notified Body for assessment. The assessment will include examination of the use of the device by lay persons and the ease of handling and understanding of the product. The Notified Body issues an EC design-examination certificate if it is satisfied. Self-test devices must have the labelling and the instructions for use in the official language of the destination Member State.

Products in Annex II fall under two lists: List A identifies reagents used for blood grouping and the diagnosis of serious diseases including HIV and hepatitis; List B identifies several other reagents for the diagnosis of serious diseases and for the self-diagnosis of blood sugar.

For IVDs in List A:

Either: the design and manufacturing procedures must comply with a quality system according to Annex IV, and the Notified Body concerned must issue a design-examination certificate;

or: the design must be shown to conform to the ERs by a type examination according to Annex V carried out by a Notified Body, and the issuing of an EC type-examination certificate. Compliance of production units to the approved design must be assured by a quality system for production according to Annex VII.

For IVDs in List B:

Either: a quality system to Annex IV must be applied but a design-examination certificate is not required;

or: the device must be given a type-examination certificate according to Annex V (as for List A products) and conformity of production must be assured by a quality system to Annex VII or by verification of batches or individual products by a Notified Body, according to Annex VI.

Current situation

None of the Member States succeeded in transposing the Directive into national law by the due date of 7 December 1999. By the end of 1st Quarter 2001, only seven countries had transposed the IVDMDD.

Commentary

The Medical Devices Directive has largely succeeded in introducing a uniform system of regulation in Europe, although many subsidiary national requirements remain. Its influence extends beyond the EC to the European Economic Area and to several countries of the former Eastern Bloc which have adopted the MDD in whole or in part. Some non-European countries are also forming or re-shaping their regulatory systems on the lines of the MDD—as discussed in Chapter 5.

Recital No. 4, which distinguishes between the regulatory requirements, which are harmonized by the Directive, and financial issues relating to healthcare provision, which are not, departs from the position of Article 3 of the Electro-medical Equipment Directive 84/539/EEC: 'Member States shall ensure that the services provided with the help of equipment meeting the requirements of this Directive are reimbursed on the same terms as the services provided with the help of equipment meeting the criteria required under the provisions in force within their territory as regards the authorized applications and minimum requirements for the equipment.' During discussion of this recital in the Commission Working Group, comment was made that this might be used to maintain national requirements and, indeed, in France, for instance, the TIPS (Tarif Interministériel des Prestations Sanitaires) authorization system for reimbursement is being continued.

The Directive has been criticized, particularly in the United States (USA 96a), for not including 'effectiveness' as a criterion for placing on the market, and Canada, which recently introduced new medical device legislation modelled largely on the MDD (Canada 98), has included an effectiveness requirement. This issue is discussed in detail in Chapter 8. More significant has been the reluctance of France to relinquish the right of the national authority to approve medical devices. This has resulted in the Law of 1 July 1998 (France 98) which requires a three-month pre-market notification to the Competent Authority in the case of Class III and many Class II devices. This move is being contested by the Commission but may lead to some tightening of the EC Directives. France has also been floating the idea of a European Medical Device Agency to address what they see as weaknesses in the operation of the Directives. So far this idea has not been given much support by other Member States but it remains a future possibility.

The use of 'Notified Bodies' as the arbiters of conformity assessment has also aroused unfavourable comment in many quarters (USA 96a) (Huriet 1996) (Kent 1996). The comments are based on the contractual relationship between the manufacturer and the Notified Body—giving rise to concerns that Notified Bodies might avoid unfavourable reports which could result in their losing paying customers—and, more generally, concerns that the Notified Bodies do not operate to a uniform standard. Certainly, the large number of Notified Bodies, spread among 15 countries, makes it difficult to ensure uniform practice. The Commission is conscious of the concerns and is using the Notified Bodies Group to share experience and has held a training course on the second edition of ISO 9001, but more effort is needed to improve confidence in the Notified Bodies. This is recognized in the Commission's paper Certif. 97/4-EN (EC 97c), the Code of Conduct for Notified Bodies (EC 98g) and the recent paper on Notified Bodies for medical devices (EC 00a). The author's view is that it would be preferable for there to be fewer Notified Bodies and for those Bodies to be notified for limited ranges of device rather than for all devices, as is generally the case. It is difficult for relatively small organizations to keep up with the latest technologies across the board and it would be a healthier situation if the Notified Bodies were to develop and maintain expertise in specific types of device. The initiation of the Notified Bodies Oversight Group is a promising move that should lead to the Competent Authorities taking a firmer grasp of their own responsibilities.

One of the difficulties experienced by Notified Bodies dealing with medical devices is the unfamiliarity of some of the tasks they have to undertake. A particular example is that of the evaluation of clinical data, discussed above. There is considerable uncertainty about just what clinical data are required for an adequate evaluation to be made and about how the evaluation is to be carried out. Notified Bodies do not generally have clinically-trained staff and have to engage clinicians as sub-contractors; clinicians are generally unfamiliar with regulatory situations. The difficulties are emphasized in the cases of essential requirements 1 and 6 which demand risk/benefit analyses which may be often made in respect of individual patients but are novel as regulatory tools. The Notified Bodies are unable to exploit the New Approach principle of compliance with harmonized standards as no standards exist, or are in preparation, for these essential requirements.

The post-market requirements of the MDD, i.e. adverse event reporting and continuous surveillance of behaviour in service, can make important contributions to the continuing safety of medical devices if properly carried out. There are second-order checks (e.g. user reporting) on adverse events but no indication of what is an adequate system of post-market surveillance. The concept may be too new to be more than an aspiration at this stage but, clearly, at some time guidance will have to be issued on this subject. It is also apparent that no one approach will be suitable for all types of device (see Chapter 9 for further discussion of post-market controls).

The problems being experienced with the application of the MDD may be merely teething troubles, as suggested by Pirovano (1997), but, even if they prove to be more permanent, they are problems of application rather than design and are primarily associated with the use of third parties in the conformity assessment process. An assessment by Kent (1998) concludes (despite his earlier criticisms (Kent 1996)) that 'the accomplishments have been impressive'. Australia, Canada and possibly other countries introducing legislation based on the MDD are restricting conformity assessment or decision making to the regulatory authority, and it may be that greater participation by Competent Authorities in the activities carried out by Notified Bodies in Europe will be necessary. It remains the author's opinion that the MDD is currently the most appropriate form of regulation for medical devices in use, and that it forms the basis of a future global regulatory system as discussed in Chapter 10. Many of the (minor) difficulties experienced with the application of the Directive are expected to be ironed out by the review now under way.

Chapter 4

The current situation: Regulations in USA and Japan—a comparison with the Medical Devices Directive

The USA and Japan have relatively long-standing medical device regulations which carry considerable weight because of the size of these two countries both as producers and consumers of medical devices. It is instructive, therefore, to examine their regulations and compare them with the Medical Devices Directive.

US Regulations

A brief, but useful, introduction to the US regulatory scheme is given by Samuel (1994). This is updated by Holstein and Wilson (1997).

The USA has had legislation governing food and drugs since 1906 and this was extended to cover medical devices by the Federal Food, Drug and Cosmetic Act of 1938 (FDCA) (USA 38). This allowed the Food and Drug Administration to take action against devices deemed to be 'adulterated' or 'misbranded'. These two terms, which are appropriate to food and drugs, have had to be given specific definitions and interpretations in order to make them applicable to devices (Samuel 1994).

The first detailed controls on products including medical devices were introduced in the Radiation Control for Health and Safety Act of 1968 (USA 68). This was followed by the Medical Device Amendments of 1976 (USA 76a) which introduced a new, comprehensive, regime for medical devices which has had a major effect on medical device safety and which established features of device regulation which have become widely accepted. This regime was modified by the Safe Medical Devices Act of 1990 (USA 90a), the Medical Device Amendments of 1992 (USA 92) and the Food and Drug Administration Modernization Act of 1997 (USA 97a). All these Acts are now incorporated in Chapter V of the Federal Food, Drug and Cosmetic Act.

72

The Medical Device Amendments of 1976

The 1976 Amendments extended the responsibilities of the Food and Drug Administration (FDA) to include:

– approving the entry of new medical devices into the market
– monitoring the compliance of medical device manufacturers with FDA laws and regulations
– requiring information about the behaviour of devices in service.

The Amendments applied to medical devices as defined in Section 201(h):

'The term "device" means an instrument, apparatus, implement, machine, contrivance, implant, in vitro reagent, or other similar or related article, including any component, part or accessory, which is
(1) recognized in the official National Formulary, or the United States Pharmacopeia, or any supplement to them
(2) intended for use in diagnosis of disease or other conditions, or in the cure, mitigation, treatment or prevention of disease, in man or other animals, or
(3) intended to affect the structure or any function of the body of man or other animals, and
which does not achieve its *primary* intended purposes through chemical action within or on the body of man or other animals and which is not dependent upon being metabolized for the achievement of any of its principal intended purposes.'

The new regulations introduced both pre-market and post-market controls and were administered by the Bureau of Medical Devices until the Center for Devices and Radiological Health was formed in 1982.

Pre-market Controls

In order to approve the entry of new devices into the market, devices were placed into one of three classes according to their perceived degree of risk (Section 513a):

• Class I (low risk) devices are those for which the 'general regulatory controls'[1] are sufficient to provide reasonable assurance of the safety and effectiveness of the device.
• Class II (medium risk) devices are those for which compliance with performance standards is necessary in addition to the general controls.

---

[1] The general regulatory controls are: prohibitions against adulteration or misbranding (Sections 501 and 502); registration of manufacturers and distributors, and product notification (Section 510); power to ban devices (Section 516) and to notify persons at risk (Section 518); reporting of adverse events (Section 519) and compliance with Good Manufacturing Practice requirements (Section 520).

- Class III (high risk) devices are those for which compliance with performance standards and general controls is not sufficient to assure their
safety and effectiveness *and* devices that are 'for a use in supporting or
sustaining human life or for a use which is of substantial importance in
preventing impairment of human health, or present a potential unreasonable risk of illness or injury.'

The classification of all devices was carried out by several panels established
according to Section 513(b). The results of the panels' deliberations were
published in the Federal Register.

Devices that were already in commercial distribution on 28 May 1976,
when the Medical Device Amendments went into effect, were allowed to
remain on the market ('grandfathered') subject to a possible call by the
FDA for a premarket approval. All devices introduced to commercial
distribution after 28 May 1976 had to be notified to the FDA under Section
510(k).

Section 513(f) stated that any notified device was to be classified in Class
III unless it was 'substantially equivalent' to a device on the market before
28 May 1976, in which case it was allowed to enter the market; if it was
not substantially equivalent to a pre-Amendment device, then a premarket
approval (PMA) was required. The FDA was required to respond to a
510(k) notification within 90 days.

'Substantial equivalence' means that a device has the same intended use
and technical characteristics as a pre-Amendment device, or has the same
intended use and different technical characteristics but is as safe and effective
as the pre-Amendment device. The FDA has issued guidance on the
assessment of substantial equivalence, and on the assessment of safety and
effectiveness of many types of medical device for use in both 510(k) notifications and PMA applications.

The PMA process is described in Section 515 of the Medical Device
Amendments. It requires the submission of full information, including the
results of a properly conducted clinical investigation, and, normally, one
or more examples of the device, to the FDA. The FDA is required to issue
a decision within 180 days.

The process of producing and establishing performance standards is
described in Section 514. However, no performance standards have been
issued by the FDA and the routes to market under the Medical Device
Amendments have been confined to the 510(k) premarket notification (and
the determination of substantial equivalence) and the premarket approval
(PMA) process.[2]

---

[2] The FDA Modernization Act of 1997 (USA 97a) introduced a provision for a manufacturer's
declaration of conformity with 'recognized standards' to be submitted in support of a 510(k) or
PMA application.

Certain special provisions associated with the pre-market controls included the following:

Custom Devices: Section 520(b) allows 'custom devices', i.e. devices made to the order of an individual physician or dentist for use by a named patient, and not made generally available, to be exempted from the provisions of Section 514 (performance standards) and Section 515 (premarket approval).

Investigational Device Exemption: Section 520(g) allows exemption from any of the requirements of the Medical Device Amendments 'to permit the investigational use of such devices by experts qualified by scientific training and experience to investigate the safety and effectiveness of such devices'.

An application for such an exemption (an 'IDE') must include sufficient information about pre-clinical testing of the device to justify its use on humans, a clinical trial protocol, approval of the local institutional review board (equivalent to an ethics committee) and informed consent of the patients involved.

Export of unapproved devices: Section 801(e) prohibited the export of devices which did not comply with Section 514 (performance standards) or 515 (PMA) unless the device:

- accorded to the specifications of the foreign purchaser
- was not in conflict with the laws of the country to which it was intended to export
- had the approval of the country to which it was intended to export, and
- was labelled as intended for export.

This provision resulted in the identification of named officials in receiving countries who were recognized by the FDA as competent to give the assurances required under this Section. Manufacturers wishing to export a device while waiting for the completion of the FDA's approval process had to obtain a letter from such an official in each country to which they wished to export confirming that all the conditions of Section 801(e) were met, then submit it to the FDA with a request for permission to export.

Post-market controls

The regulatory provisions did not end with the permission to market the device. Throughout the life of the device, the manufacturer was required to maintain certain controls in place. The most important of these were:

Good Manufacturing Practice: Pursuant to Section 520(f) of the Medical Device Amendments, the FDA published a Good Manufacturing Practice (GMP) Regulation in the Federal Register of 21 July 1978 (USA 78). This

Regulation imposed requirements relating to 'methods used in, and the facilities and controls used for, the manufacture, packing, storage and installation of all finished devices intended for human use.'

This Regulation was the first to impose what are now called quality system requirements on the manufacturers of medical devices and it had a profound effect on such manufacturers, not only in the United States but in every country from which devices were exported to the United States. Checks on compliance with the GMP requirements were made by inspectors from the FDA field force and were carried out after the approval of the device. Visits to manufacturers in the USA were generally unannounced but visits to overseas manufacturers were made by appointment. (This was not a matter of principle but was related to the time and expense involved in making overseas visits). Serious deviations from GMP, or deviations that were not rectified quickly, resulted in the banning of the device or, in the case of overseas manufacturers, possibly the banning of all devices made by the transgressing manufacturer (again, because of the difficulty of making repeated visits overseas to check that the remedial action had been carried out). (See also Chapter 6.)

Medical Device Reporting: Section 519 of the Medical Device Amendments provides for the issuing of regulations which would require manufacturers to maintain records, make reports, and provide such information as might reasonably be required 'to assure that such device is not adulterated or misbranded and to otherwise assure its safety and effectiveness.'

The Medical Device Reporting (MDR) Regulation was published in the Federal Register of 14 September 1984 (USA 84). The core of this Regulation is:

'FDA is requiring a device manufacturer or importer to report to FDA whenever the manufacturer or importer receives or otherwise becomes aware of information that reasonably suggests that one of its marketed devices:
(1) may have caused or contributed to a death or serious injury, or
(2) has malfunctioned and that the device or any other device marketed by the manufacturer or importer would be likely to cause or contribute to a death or serious injury if the malfunction were to recur.

In addition, a device importer is required to establish and to maintain a complaint file and to permit any authorized FDA employee at all reasonable times to have access to and to copy and verify the records contained in this file.'

The time scales for reporting were:

Death or serious injury	telephone report within 5 days, written report within 15 days
Malfunction	written report within 15 days.

This was another innovation of the Medical Device Amendments which has proved to be a key factor in assuring the public health and has been adopted in most other administrations (see Chapter 9).

The Safe Medical Devices Act of 1990 and the Medical Devices Amendments of 1992

The Safe Medical Devices Act of 1990 (USA 90a) introduced several substantial modifications to the 1976 Amendments and these were further refined by the 1992 Medical Device Amendments (USA 92). These were:

Class II devices

'Congress recognized that, as written in the 1976 Medical Device Amendments, the procedures for establishing performance standards were unworkable' (Holstein and Wilson (1997) p. 215). Consequently the SMDA changed the requirements for Class II devices, making them subject to 'special controls' (as opposed to performance standards) in addition to general controls. These special controls include 'the promulgation of performance standards, postmarket surveillance, patient registries, development and dissemination of guidelines (including guidelines for the submission of clinical data in premarket notification submissions in accordance with section 510(k)), recommendations and other appropriate actions as the Secretary [of Health and Human Services] deems necessary....' (Section 513(a)(1)(B).

These changes recognized the fact that the FDA had not produced any performance standards and that premarket examination of medical devices could not always guarantee their lifetime safety so that other techniques to improve the protection of patients had to be considered.

510(k) notifications

Section 513 was modified to include the concept of a 'predicate device', i.e. a device currently legally marketed. This recognized the passage of time and the incremental changes made in many medical devices since 1976. Substantial equivalence to a predicate device in a 510(k) notification could ensure the approval of a device. However, this apparent relaxation in the requirements was balanced by the changes in Section 513, mentioned above, which allowed the FDA to require the submission of clinical data in 510(k) notifications.

Post-market controls

A number of changes were made to strengthen the part played by post-market actions in the control of medical device safety:

- The adverse event reporting obligation of Section 519 was extended to 'device user facilities' (hospitals, nursing homes, ambulatory surgical facilities, and outpatient diagnostic and treatment facilities) and to distributors; the 1992 Amendments clarified the reporting requirements and introduced a single definition of a reportable event: 'an injury or illness that is: (1) life-threatening; (2) results in permanent impairment of a body function or permanent damage to a body structure; or (3) necessitates medical or surgical intervention to preclude permanent impairment of a body function or permanent damage to a body structure.' All these changes were embodied in a Final Rule published in the Federal Register of 11 December 1995 (USA 95a).
- Section 519(e) was added requiring a manufacturer of '(A) a permanently implantable device, or (B) a life sustaining or life supporting device used outside a device user facility' to adopt a method of device tracking. This provision could be extended to other designated devices. The implementation of this requirement was delayed by the 1992 Amendments.
- Section 522 introduced a requirement on manufacturers to conduct post-market surveillance on: permanent implants the failure of which might cause serious adverse health consequences or death; devices intended for a use in supporting or sustaining human life; or devices which potentially present serious risks to human health.

The FDA Modernization Act of 1997

The Food and Drug Administration Modernization Act (FDAMA) (USA 97a) was signed into law on 21 November 1997. Some provisions became effective on 19 February 1998; some provisions were staggered; and others required notice-and-comment rulemaking to become effective.

Pilot and Waldmann (1998) and Barlow (1999) describe the background to FDAMA. Demand for some changes in the way medical devices were regulated by the Food and Drug Administration (FDA) began soon after the enactment of the Safe Medical Devices Act of 1990 and was stimulated by the publication of the Congressional Report *Less Than the Sum of its Parts* (USA 93a) and a report by the Wilkerson Group (1995) which criticized the time taken by the FDA to allow new medical devices to reach the market. Proposals for change were concentrated on introducing the use of third parties in the approval process, as in the European medical device Directives, and on limiting the scope of effectiveness studies (see Chapter 8). FDA reaction to these proposals was given in a paper issued in March 1995 (USA 95b). In the event, these issues took their places among other, equally significant, developments, the most important of which were (see also Davis (1998)):

- FDA will be more open to meet with manufacturers to discuss clinical studies and PMAs (Sections 201, 205 and 209).
- Under defined conditions, changes may be made to device manufacture and design, or clinical study protocols without additional FDA approval (Sections 201 and 205).
- FDA may recognize all or part of a national or international standard for use in PMA or 510(k) submissions. Manufacturers may offer declarations of conformity to recognized standards in such submissions (Section 204).
- FDA is required to consider the least burdensome means of evaluating effectiveness for a PMA, and the least burdensome method of establishing substantial equivalence for a 510(k) submission. In both cases, FDA must consider if postmarket controls can reduce the data or time required for clearance (Section 205).
- A 510(k) submission is not required for a Class I device unless it presents a potential unreasonable risk or is of importance in preventing impairment of human health. Certain Class II devices are to be exempted from the 510(k) process (Section 206).
- Third parties ('accredited persons') will be allowed to conduct initial 510(k) reviews on a large group of Class II devices (Section 210).
- Mandatory tracking and postmarket surveillance requirements are replaced by the requirement for a determination by the FDA that tracking or postmarket surveillance is needed (Sections 211 and 212).
- FDA is to ensure that the Quality Systems Regulation conforms to international standards 'to the extent practicable', and is to meet with other countries for the purpose of harmonizing regulatory requirements.

Progress on implementing FDAMA is recorded on the website:

http://www.fda.gov/po/modact97.html

and a brief account of some significant events is given here:

The Federal Register of 25 February 1998 (USA 98a) listed 174 standards 'recognized' in terms of Section 204 of the Act and this number has steadily increased, reaching 501 by end-2000. The Federal Register of 21 January 1998 and 2 February 1998 listed Class I and Class II devices exempted from regulatory procedures under Section 206. Both these lists have been updated and will continue to be updated from time to time. They are maintained on the CDRH website at http://www.fda.gov.cdrh. Criteria for the accreditation of third parties were published in May 1998 (USA 98b) and a guidance document on the third party program was issued on 30 October 1998 and updated in February 2001 (USA 01a). At end-2000 the CDRH website listed 12 bodies classed as 'accredited persons'. A Final Rule on medical device reporting was published in January 2000 (USA 00a) and a Proposed Rule on tracking (see Chapter 9) was published in April 2000 (USA 00b).

The Quality Systems Regulation

The SMDA of 1990 extended Section 520(f) to cover controls on 'pre-production design validation (including a process to assess the performance of a device but not including an evaluation of the safety or effectiveness of a device)' in addition to the existing controls on the manufacture of medical devices.

This necessitated a substantial change to the Good Manufacturing Practice Regulation (USA 78) and the opportunity was taken to revise the regulation thoroughly and to bring it into close alignment with the international standard on quality systems for design and manufacture ISO 9001 (ISO 94a). There were lengthy periods of public consultation and consideration by the Global Harmonization Task Force[3] before the revised Regulation was published in the Federal Register of 7 October 1996 (USA 96b). It has since been supplemented by design control guidance published in March 1997 (see Chapter 6).

The Radiation Control for Health and Safety Act of 1968

The Radiation Control for Health and Safety Act (USA 68) was enacted in 1968 to protect the public from unnecessary exposure to radiation of all kinds from electronic products including, but not confined to, certain medical devices. The Act was administered by the Bureau of Radiological Health until 1982 when the Bureau of Radiological Health was merged with the Bureau of Medical Devices to form the Center for Devices and Radiological Health.

The electronic products covered by the Act include all products capable of emitting ionizing or non-ionizing radiation, such as microwave ovens, sun lamps, infrared heaters, ultrasonic cleaners and medical devices such as X-ray equipment, diathermy units, particle accelerators, ultrasonic devices and lasers. The Act provides for devices to be certified by the manufacturer as complying with applicable performance standards issued by the FDA. Standards have been published from time to time in the Federal Register and are compiled in Title 21, Code of Federal Regulations, Parts 1020 to 1050.

In 1990, the Radiation Control for Health and Safety Act was recodified as the Electronic Product Radiation Control (EPRC) provisions of the Federal Food, Drug and Cosmetic Act. All the requirements of this Act,

---

[3] The Global Harmonization Task Force is an informal body of regulators and industry from several major countries which exists to reduce the differences in medical device regulations world-wide. See Chapter 10.

discussed above, apply to radiation-emitting devices as well as the radiation control requirements.[4]

Advice and Guidance

As in other countries, advice/guidance is issued frequently to medical device manufacturers and to FDA staff. A Quarterly List of Guidance Documents at the FDA is published in the Federal Register and maintained on the website: http://www.fda.gov/OHRMS/DOCKETS/98fr/031400f.txt.

A comparison of US requirements with those of the Medical Devices Directive

Comparisons between the US and EC regulatory systems have been published by the US General Accounting Office (USA 96a), Hastings (1996) and, more recently, by Chai (2000). The two earlier studies were written in defence of the US system and even Chai has carried over some of their views. This chapter, although written by a European who was involved in the development of the EC Medical Devices Directive, attempts to take a more balanced approach.

The US Medical Device Amendments of 1976 introduced several concepts which were adopted by the European legislators and incorporated (with some modification in some cases) in the MDD. Examples are:

- Classification of devices, according to perceived risk, with different conformity assessment procedures for each Class. It was originally intended that the MDD should have three classes corresponding to those of the US MDA. However, the idea of classification panels, as used in the US, was unacceptable to the European Commission because of the time, cost and bureaucracy involved, and the exercise of developing rules by which the manufacturer could classify devices led to the introduction of four classes in the MDD. Despite the impossibility of matching three classes against four, there is a general correspondence in the gradation of medical devices in the two systems.
- Good Manufacturing Practice. The Medical Devices Directive incorporated quality system requirements which were effectively an updated version of the US GMP Regulation of 1978 but were at three different levels according to the three levels of the ISO 9000 series. The revision of the US Regulation in 1996 brought it into close alignment with

---

[4] For a fuller description of the regulations governing medical radiation safety in the USA see the Chapter by Bennett and Dormer (1997) listed in the bibliography.

Annex II of the MDD. Although the requirements are very similar, they are used in different ways: in the US, the requirements are mandatory for all Class II and Class III devices as well as certain Class I devices, whereas in the MDD compliance with any of the quality system Annexes is optional. Furthermore, in the MDD compliance with quality systems is a pre-market requirement whereas that is not the case in the US (although the FDA are now refusing to accept 510(k) or PMA submissions unless the manufacturer has had a recent satisfactory inspection).

- Vigilance. The adverse event reporting and post-market surveillance requirements of the MDD are drawn from the US system and are essentially the same.
- Investigational and custom devices. The exemptions from the regulations, under defined conditions, for investigational and custom devices are very similar in the two systems.

The differences between the US and European models are fundamental. The first of these is in the prime object of the regulations:

- In the US regulations devices are to be safe and effective, where the effectiveness of a device is to be determined 'on the basis of well-controlled investigations, including clinical investigations where appropriate, by experts qualified by training and experience to evaluate the effectiveness of the device, from which investigations it can fairly and responsibly be concluded by qualified experts that the device will have the effect it purports or is represented to have under the conditions of use prescribed, recommended, or suggested in the labelling of the device.' (Section 513(a)(3)(A) Medical Device Amendments.) The Medical Devices Directive contains no mention of effectiveness or of an equivalent term, the key requirement being that devices 'will not compromise the clinical condition or the safety of patients, or the safety and health of users or, where applicable, other persons, *provided that any risks which may be associated with their use constitute acceptable risks when weighed against the benefits to the patient* (author's italics) and are compatible with a high level of protection of health and safety.' (MDD Annex I paragraph 1.) This identifies special cases when the benefits offered by a device (which may be construed as its 'effectiveness') have to be determined. Devices are also required to 'achieve the performances intended by the manufacturer.' (See Chapter 8 for a more detailed discussion.)
- Reference to standards. The 'deemed to satisfy' condition offered by harmonized standards is a key feature of all New Approach Directives (EC 85b). Although the US Medical Device Amendments specified compliance with performance standards for Class II devices, no standards were promulgated and assessments under 510(k) or PMA were conducted against the general 'substantial equivalence' and 'safety and

effectiveness' requirements. The Food and Drug Administration Modernization Act of 1997 introduced the concept of 'recognized standards' to be used in product reviews. The importance and usefulness of this provision remains to be seen. (See also Chapter 7 and Chapter 10.)

– Flexibility in conformity assessment. The MDD offers some degree of manufacturer choice in the conformity assessment method to be used for devices in all classes other than Class I. No such flexibility exists in the US system.

– Conformity assessment by third parties. To date 62 Notified Bodies have been identified in Europe. Any of these can be chosen by the manufacturer to carry out the conformity assessment procedures leading to the affixing of the CE marking. The majority of these bodies are private-sector organizations. The US laws allow only the Food and Drug Administration to determine the acceptability of medical devices although a 'pilot program' was introduced in 1996 (USA 96c) under which some private-sector bodies may carry out 510(k) examinations of selected devices and report to the FDA who will make the decision. This provision was extended by the FDAMA for a period of four or five years, depending on the way the program was taken up. By November 1998 FDA had accredited 13 organizations to review 510(k)s and terminated the pilot program. The takeup was poor; in the first 17 months only 28 companies used third parties to review 54 510(k) submissions and in July 2000 FDA published a notice in the Federal Register (USA 00c) revising the guidance on the third party program, expanding the list of eligible devices, and seeking further accredited persons in an attempt to increase the use of the program. At the end of 1st Quarter 2001, 12 accredited persons (organizations) were listed on the FDA website.

A Mutual Recognition Agreement (MRA) between the EC and the USA was signed in May 1998 (Dickinson (1997), EC 98i, USA 98c) and came into force on 1 December 1998 under which designated bodies in the EC will be allowed to carry out 510(k) examinations of a range of devices listed in an Appendix to the MRA, and designated bodies in the USA will be able to carry out the conformity assessment procedures of the Medical Devices Directive and the Directive on Active Implantable Medical Devices. A three-year transitional period is allowed for confidence building. A progress note from the European Commission (EC 00c) stated that potential conformity assessment bodies had been identified by both parties and observed audits and training workshops had begun. A chart of implementation activities and progress is given on the website http://www.fda.gov/cdrh/mra/impchart.html.

Successful operation of this MRA would certainly help the cause of world-wide recognition of approvals, as discussed in Chapters 10 and 11.

Medical Device Regulation in Japan

Medical devices are regulated in Japan under the Pharmaceutical Affairs Law (PAL), administered by the Ministry of Health and Welfare (MHW). This was originally promulgated in 1943 and was amended to include medical devices in 1948 although no specific provisions were made for devices. The definition of a medical device introduced then remains the same (Article 2.4):[5]

> 'The term "medical device" in this Law refers to equipment or instruments intended for use in the diagnosis, cure or prevention of disease in man or animals, or intended to affect the structure or functions of the body of man or animals, and which are designated by cabinet order.'

In vitro diagnostic reagents, which are not equipment or instruments, are not medical devices and are regulated as quasi-drugs.

In 1960, licence and approval requirements were added (Japan 60), but the major developments were in 1979 when Regulations on Handling and Safety of Medicine were published and requirements for clinical investigation and adverse event reporting were added. In 1987 a Good Manufacturing Practice Regulation (Japan 87) was published and in 1994 the medical device provisions of the PAL were brought up-to-date (Japan 94a).

Latest developments: In 1998 a new classification system for medical devices, based on the draft proposals of the Global Harmonization Task Force (see Chapter 10) was announced (Japan 98). This introduced four classes based on rules, with associated conformity assessment procedures. The full effects of this new system and its interlacing with the pre-existing PAL, Enforcement Ordinance and Enforcement Regulations are not yet clear and the following discussion is based on currently available documents.

The Pharmaceutical Affairs Law

The purpose of the Pharmaceutical Affairs Law is to 'control and regulate matters related to drugs, quasi-drugs, cosmetics and medical devices to assure their quality, efficacy and safety'. It consists of the following chapters:

1. General Provisions (Articles 1, 2)
2. Pharmaceutical Affairs Council (Articles 2–4)
3. Pharmacies (Articles 5–11)
4. Manufacturer and Importer of Drugs, etc. (Articles 12–23)

---

[5] Details of the Japanese laws, regulations and ordinances affecting medical devices are based on a translation of the Pharmaceutical Affairs Law: Extracts of Parts Affecting Medical Devices (Japan), Edited by The Japan Federation of Medical Devices Associations (JFMDA) and published by Yakuji Nippo Ltd. Tokyo, Second Edition 1998 (see also Ohashi (1998)).

5. Sale of Drugs and Sale and Lease of Medical Devices (Articles 24–40)
6. Standards and Tests for Drugs, etc. (Articles 41–43)
7. Handling of Drugs, etc. (Articles 44–65)
8. Advertising of Drugs, etc. (Articles 66–68)
9. Supervision (Articles 69–77.2)
10. Miscellaneous Provisions (Articles 77.3–83)
11. Penal Provisions (Articles 84–89)

and is supported by an Enforcement Ordinance, originally published in 1961 (Japan 61a) and revised in 1995 (Japan 95a) and Enforcement Regulations originally issued in 1961 (Japan 61b) and revised in 1995 (Japan 95b).

Provisions affecting medical devices

As may be seen from the chapter headings, the PAL and its associated documents is difficult to follow because it contains requirements related to medical devices, medicinal products and cosmetics—often within the same Article. Furthermore, announcements from the MHW are often used to promulgate new or changed requirements which may not find their way into the PAL for some years, if at all. Foreign manufacturers wishing to export to Japan can find it difficult to determine the specific approval procedures for a particular device.

The Japanese legislation has a style of its own but includes several features found in the European or US systems. It imposes pre-market controls on both products and manufacturers and, in the recent revisions, places requirements on manufacturers to monitor in various ways the behaviour of devices in service. One individual feature of the Japanese system is the inclusion of requirements specific to importers and imported medical devices.

Manufacturing licence: Article 12 of the PAL stipulates that a licence ('kyoka') is required for the manufacture of medical devices. To obtain the licence the plant must comply with MHW standards (including quality assurance or GMP) (Article 13) and each device to be manufactured must have its own approval ('shonin') (Article 14). Article 1–2 of the Enforcement Ordinance, and the attached Table 2, identifies a number of devices which are exempted from the licence standards of Article 13. The MHW may specify particular test or inspection methods to be used by manufacturers (Article 16) and the manufacturer must employ a 'responsible technician'[6] at each plant 'to supervise in practice the manufacture of drugs, quasi-drugs or

---

[6] The qualifications of responsible technicians are described in Article 24 of the Enforcement Regulations. For devices which need approval, they are a university education in an appropriate subject or equivalent training and experience. For devices exempt from approval, high school education or equivalent training and experience is needed.

medical devices' (Article 17). The application for a licence to manufacture is made to the government of the prefecture in which the plant is situated.

Device approval: Article 14 of the PAL requires all devices to be approved and outlines the information to be submitted in an application for approval. More details of the information required is given in Article 18.3 of the Enforcement Regulations. This is a general requirement but there are some variations:

–   Article 18 of the Enforcement Regulations lists 82 types of simple device which are exempted from Article 14, i.e. do not need approval, plus 126 devices for which compliance with a Japanese Industrial Standard (JIS) gives exemption from Article 14.
–   Article 77.5 of the PAL states that certain devices, implanted in the human body and intended to be used outside medical facilities, are subject to additional requirements, viz. tracking of their locations and the maintenance of records. The specific devices subject to these additional requirements are listed in Article 64.6 of the Enforcement Regulations as:
    –   implantable cardiac pacemakers and their leads
    –   implantable defibrillators and their leads
    –   artificial cardiac valves
    –   artificial cardiac valve rings
    –   artificial blood vessels (limited to use for the coronary artery, thoracic aorta and abdominal aorta).

There are two other specific groups in the Japanese system:

–   devices for home use require approval;
–   'new medical devices' defined in Article 14.4 of the PAL as 'medical devices which are designated by the Minister at the time of approving manufacture as being significantly different from those already approved for manufacture or import . . . in terms of the structure, method of use, indications, effects, performance' have to be re-examined after a period between four and seven years. The re-examination is to be based on the results obtained during this period of use. Follow-up and reporting procedures are described in Article 21.4 of the Enforcement Regulations.

Conformity assessment: Applications for manufacturing licences or for the approval of devices are made to the governor of the prefecture in which the manufacturing plant (or the 'in-country caretaker'—see section on imported devices) is based (Article 20 of the PAL). The application is forwarded to the MHW who issue the approvals. The data to be included in the application for a manufacturing licence are specified in Article 14 of the Enforcement Regulations. The data to be included in an application for product approval are listed in Article 18-3 of the Enforcement Regulations as:

A. Data concerning the origin of the medical device, the background of its discovery and the conditions of use in foreign countries
B. Data concerning the physiochemical properties, standards and test methods
C. Data concerning stability
D. Data concerning safety including electrical safety, biological safety, and safety with respect to radiation
E. Data concerning performance
F. Data concerning the results of clinical trials.

Article 14 paragraph 4 of the PAL states that the approval method shall include an examination of the equivalence of the item under consideration to an item already approved for manufacture or import. Article 14.3 of the PAL allows the Minister to delegate all or part of this examination to a 'designated investigation organization' (DIO). Articles 28-2 to 28-15 of the Enforcement Regulations describe the requirements for establishment as a DIO and the method of investigation that a DIO must use—stipulated as comparison with already approved devices. To date the Japan Association for the Advancement of Medical Equipment (JAAME) is the only DIO.

Article 42 of the PAL empowers the MHW to 'lay down the necessary standards related to the properties, quality, efficiency, etc. of ... medical devices' 'when it is indispensable for safeguarding public health'. Article 43 states that in such cases the designated devices may not be sold, leased, given, stored or exhibited unless they have passed the tests carried out by a designated person (organization) and that the designations will be made by cabinet order. Hirai and Nagao (1994) refer to 20 standards in force at that time, including: syringe needles, syringes, transfusion sets, blood donor sets, artificial blood vessels, contact lenses, plastic sutures, latex condoms, sanitary tampons, artificial heart valves, cardiac pacemakers and X-ray apparatus.

Sales notification: Article 39 of the PAL requires 'any person who intends to sell or lease the medical devices designated by the Minister' to notify the governor of the prefecture in which the business office is located. Article 41 of the Enforcement Regulations refers to attached table 2 which lists 47 categories of designated devices, and Article 42 identifies the information to be notified. Article 42.2 imposes quality assurance requirements on sellers and leasers of medical devices.

Imported devices: Article 22 of the PAL requires importers of medical devices to obtain an import licence and Article 23 states that all the articles governing the obtaining of a manufacturing licence apply to the obtaining of an import licence. Imported products must be approved in the same way as domestically-produced medical devices (Article 19.2). Article 19.2, paragraph 3 introduces a requirement that 'in order to take the necessary measures to prevent public health hazard by... medical devices approved, a person who seeks approval... shall appoint, at the time of approval, a person who conforms to the standards specified in MHW ordinance from among persons domiciled in Japan.' Such a person is designated in Article 19.3 as the 'in-country caretaker'. The in-country caretaker carries out all the approval activities on behalf of a foreign manufacturer.

Adverse reaction reporting: Article 62-2 of the Enforcement Regulations requires the manufacturer or the importer or the in-country caretaker to report adverse reactions to the MHW on the following basis:

'(1) Within 15 days if the information is related to the following.
A. Unexpectable (*sic*) death, disablement or other related cases suspected to be caused by deficiencies in the medical devices being used in Japan. But, this is also applicable to those devices being used abroad, if they are similar...
B. Deficiencies of medical devices, which might lead to the cases mentioned in A.

(2) Within 30 days if the information is related to the following, but excluding the preceding item.
A. Death, disablement or other related cases which are suspected to be caused by deficiencies in the medical devices, and which are almost incurable or found serious by the physician or dentist in charge.
B. Disorders which are suspected to be caused by deficiencies in the medical device, and which are found not insignificant by the physician or dentist in charge. The said devices have not any precautions for use in the accompanying documents or on the container or package (*sic*).
C. Deficiencies of medical devices, which might lead to the cases mentioned in A or B.
D. Research reports about the following: possible deficiencies in medical devices which might have serious effect on human health; remarkable changes in the trend of incidence, such as number of cases, frequency, conditions of incidence, etc. of deficiencies which might have some effect on human health; or the lack of the approved indications, effects or properties.'

Clinical trials: Article 80-2 of the PAL stipulates that clinical trials on any device that requires approval must be notified to the MHW which may prohibit the trial. Articles 67–70 of the Enforcement Regulations give the detailed requirements including, for foreign manufacturers, the appointment of an 'in-country clinical trial administrator'.

Good Manufacturing Practice: A Good Manufacturing Regulation was introduced in 1987 (Japan 87) and was replaced by a Quality Assurance

System for Medical Devices published in 1994 (Japan 94b). This standard applies to all manufacturers/importers given a licence under Article 12 of the PAB. The 1994 version was extensively modified as a result of comments from the Global Harmonization Task Force and is fairly well aligned with the international standard ISO 9001 (ISO 94a). Inspectors from MHW check domestic manufacturers for compliance but rarely, if ever, go overseas. Guidance to prefectural authorities on the application of the quality assurance standard to imported medical devices is contained in Yaku-Hatsu No. 380, 19 April 1993, and Yaku-Kan No. 19, 31 March 1994. The latter document states that the following procedures are acceptable:

'(1) For verification, importers shall use the GMP certificate issued by the government of the exporting country.
(2) For verification, importers shall use the GMP audit records or reports issued by the government of the exporting country.
(3) For verification, importers shall use the information (all the information specified in items (a) and (b) below) concerning quality assurance of medical devices, provided by the manufacturers in the exporting country.
   (a) Information concerning the GMP standards of the exporting country, or equivalent standards for quality assurance of medical devices.
   (b) A statement by manufacturer in the exporting country that said medical devices conform to the GMP standards or equivalent standards for quality assurance of medical devices in the exporting country. The statement shall be signed by the director in charge of the manufacturing country.'

It goes on to say that the European standards EN 46001 (CEN 97a) and EN 46002 (CEN 97b) are acceptable standards for the application of item (3).

Miscellaneous issues: The PAL, the Enforcement Ordinance and the Enforcement Regulations specify a number of other issues, including:

— labelling (Article 63 of the PAL, Article 61 of the Enforcement Regulations)
— selling or leasing of devices (Articles 39 and 40 of the PAL, Article 41 to 45-2 of the Enforcement Regulations)
— advertising restrictions (Articles 66 to 68 of the PAL)
— installation control (Articles 23-2 of the Enforcement Regulations)
— repair of medical devices (Articles 23-3 to 23-5 of the Enforcement Regulations)
— the Central Pharmaceutical Affairs Council (Article 3 of the PAL).

A comparison of Japanese requirements with those of the Medical Devices Directive

The 1994 medical device requirements of the Pharmaceutical Affairs Law has certain elements which correspond to those of the MDD. Examples are:

- Classification of devices, leading to variations in the conformity assessment procedure. However, this classification has not been explicit until now and has consisted of the majority of devices being subject to general requirements, with small numbers being either exempted or subject to additional requirements as strictly determined by the Ministry of Health and Welfare. The separate designations of devices for home use and 'new medical devices' are not found in the European legislation. The new classification system announced in 1998 (Japan 98) is a four-class system which resembles that of the European Medical Devices Directive, although the conformity assessment procedures are somewhat different (Tsukamoto 1999). Nevertheless, this move appears to represent a start to a process of bringing Japanese legislation closer to that of the EC.
- The delegation of certain conformity assessment procedures to a non-governmental third party. In practice, this meant that almost all devices were examined by JAAME (the role of JAAME in the new system is not yet clear), but, unlike the European system, only one third party is authorized and JAAME does not have the authority to issue approvals, this being retained by MHW.
- Quality systems. The requirement for manufacturers to comply with the Quality System Standard before obtaining a manufacturing or import licence is similar to the use of Annex II of the MDD for Class IIa and Class IIb devices. However, the mandatory nature of the Japanese quality system requirements differentiates the Japanese system from the MDD.
- Vigilance. The adverse reaction reporting requirements are very similar to those of the MDD. Furthermore, the follow-up requirements for 'new medical devices' are similar to the post-market surveillance requirements of the MDD, although these apply to all devices.
- Clinical investigations. The requirements in the PAL for clinical investigations to be notified to the MHW, the form of notification and the conditions for the conduct of clinical investigations are very similar to those of the MDD.

Significant differences between the Japanese and the European legislation are:

- The general purpose of the PAL, as expressed in Article 1 is far wider than that of the MDD: 'The purpose of this law is to control and regulate matters related to drugs, quasi-drugs, cosmetics and medical devices to assure their quality, efficacy and safety. It is also to improve public health and sanitation by taking measures to promote the research and development of the drugs and medical devices that are greatly required in medical practice.' This may be compared with the aim of the MDD which is restricted to ensuring 'that devices may be placed on the market and put into service only if they do not compromise the safety and health of patients etc.'

– The definition of a medical device is much less precise than that of the MDD and relies, in the last resort, on designation by the MHW. It does not include in vitro diagnostic reagents. There is no reference to accessories in the PAL.

– The assessment of acceptability is generally made on the basis of equivalence to already approved devices rather than compliance with stated requirements (although the new classification scheme includes compliance with designated standards for Class II). Clinical data are always required so that an assessment of effectiveness can be made but clinical trials are only mandatory for Class IV. No interpretation of the term 'efficacy' is given, but Article 14 paragraph 2 of the PAL states that approval shall not be granted if:

'(1) The drug, quasi-drug or medical device is not shown to possess the indications, effects or properties indicated in the application.

(2) The drug, quasi-drug or medical device in the application is found to have no value as a drug, quasi-drug or medical device because it has harmful action which outweighs its indications, effects and properties.'

– The need for a manufacturing licence, in addition to the approval of each device (unless exempted), is a feature not found in the MDD. Features associated with the licensing of manufacturers, such as the appointment of a 'responsible technician' are not found in the MDD.

– The specific requirements for imported devices, including the need for an import licence, are not found in the European system.

– The PAL requires sellers or leasers of certain designated devices to notify their activity to the prefectural government. As the list of designated devices appears to be a selection from the list of devices exempt from approval, this may correspond, in some way, to the MDD requirement for notification by sellers of Class I devices.

– The provisions for re-examination of 'new medical devices' and for tracking of the designated (Class 3—presumably now Class IV) devices are requirements not found in the MDD.

– In general, the highly prescriptive nature of the PAL, the Enforcement Ordinance and the Enforcement Regulations contrasts with the MDD which lays down the important features but leaves many matters of detail to the Member States.

Commentary

Some of the marked differences between the Medical Devices Directive and other major legislation, in Japan and the USA, can be linked to the fact that the Japanese and US medical device regulations form a subsidiary part of pharmaceutical regulations, whereas the European Directive was conceived and drafted independently of the (existing) Medicinal Products Directive.

The derivation from pharmaceutical legislation has resulted in the carry-over of concepts and language which are inappropriate for medical devices. Examples are the use of the terms 'adulterated' and 'misbranded' in the US legislation. The definitions of these terms given in Sections 501 and 502 of the Food, Drug and Cosmetic Act are long and complex. They include such language as:

> [A drug or device is adulterated or misbranded] 'If it consists in whole or in part of any filthy, putrid or decomposed substance...if it has been prepared, packed, or held under unsanitary conditions whereby it may have been contaminated with filth...if it bears or contains, for purposes of coloring only, a color additive which is unsafe within the meaning of section 706(a)...if...its strength differs from, or its purity or quality falls below, that which it purports or is represented to possess... If in a package form unless it bears a label containing (1) the name and place of business of the manufacturer, packer or distributor; and (2) an accurate statement of the quantity of the contents in terms of weight, measure, or numerical count.... If it is for use by man and contains any quantity of the narcotic or hypnotic substance...unless its label bears the...statement 'Warning—may be habit forming'.... Unless its labelling bears (1) adequate directions for use; and (2) such adequate warnings against use in those pathological conditions or by children where its use may be dangerous to health, or against unsafe dosage or methods or duration of administration or application...'.

The definitions include aspects specific to devices, generally relating to errors or omissions by the manufacturer with respect to the approval processes, but these appear to be contortions of reasonable requirements, which could have been expressed in straightforward language, in order to fit in with pre-existing terms more readily applied to food or drugs. As Samuel (1994) puts it 'These terms have specific definitions in the law that go far beyond common definitions. In fact, the terms are anachronisms, first written into the law in an earlier, simpler technological age long since superseded. Yet they continue on, encrusted with additional meanings designed to adapt their original scope to the infinitely more complex technologies and patterns of use that characterize the late 20th century.'

The Japanese law also incorporates to devices features more appropriate to drugs, such as 'The approvals...shall be based on the examination of the name, ingredients, quantities, structure, directions, dosage, method of use, indications, effects, performance, adverse reaction, etc. of the drug, quasi-drug, cosmetic or medical device concerned.' (Article 14 paragraph 2 PAL.) Requirements relating to the proprietors of pharmacies are applied to 'the profession of selling or leasing medical devices' by Article 40 of the PAL. The requirement for a 'responsible technician' in each manufacturing plant is probably a carry-over from the days when many drugs were made up by individual pharmacists.

Even more fundamental differences between the MDD and the Japanese and US systems almost certainly derive from the origins of the medical device legislation. These are, firstly, the inclusion of 'effectiveness/efficacy' as a criterion for approval in Japan and the USA, but not in Europe, and, secondly, the omission of detailed requirements for safety (and efficacy) in the Japanese and US legislation, compared with the listing of 'essential requirements' in the MDD together with the 'deemed to satisfy' status of 'harmonized standards'. These two subjects are discussed in detail in Chapters 7 and 8.

A further significant difference between the European system and those in Japan and USA is in the nature of the permission to market a medical device. In both Japan and USA, a formal approval is given by the government to a type of device. In the USA, approvals are published in the Federal Register and, in Japan, they are entered in a register maintained by the Ministry or the provincial government. In Europe, no formal approval is given; the manufacturer, having gone through the appropriate conformity assessment procedure for which he may have obtained certificates or reports from a Notified Body, issues a declaration of conformity in respect of *each individual device* and confirms this by the affixing of the CE marking. The use of third parties for the conformity assessment procedures has been criticized in the past by US authorities (USA 96a) but some tentative moves in this direction have been made in the past two years and if the Mutual Recognition Agreement between the EC and the USA works successfully this principle may be more readily accepted.

Despite these differences, many elements found in the three systems are so similar that they may be regarded as establishing a fundamental approach to the regulation of medical devices. These are:

- The attempt in the European and US definitions of a medical device to distinguish devices from drugs on the basis of the 'primary intended purposes' (US), 'principal intended action' (Europe) not being accomplished by chemical, pharmacological, metabolic, etc. means. The MDD definition is more advanced as it recognizes that a device may be assisted in its action by such means.
- In all three systems, the classification of devices according to perceived risk; the application of conformity assessment procedures of increasing severity according to class; the acknowledgement that devices in the

lowest class do not need approval; the need for devices in the highest class to be individually approved, taking account of clinical data.

- The application of quality systems by manufacturers as a key factor in ensuring the consistent safety of medical devices. Like many other elements of medical device regulation, this was first introduced by the USA in the Medical Device Amendments of 1976. In common both with the approach to consistent manufacture of pharmaceuticals and the development of quality systems at that time, the first US GMP Regulation (USA 78) did not include control of design, neither did the first Japanese GMP (Japan 87). Although the international standard for quality assurance, ISO 9001 (ISO 94a), covered design control, the MDD was the first major legislation to apply design control to medical devices and to make it a pre-market condition. Japan and USA revised their GMP Regulations to include design controls and to resemble closely the international standard. Japan uses their quality system as a pre-market condition and the US FDA increasingly require confirmation that quality systems are applied before processing PMA or 510(k) applications, so there is steady progress towards a consistent approach to the content and application of quality assurance in medical device regulation.

- Adverse event reporting. Mandatory adverse event reporting by manufacturers was, again, introduced in the US Medical Device Amendments of 1976 and virtually the same requirement is found in the Japanese Enforcement Regulations and the MDD.

- Post-market surveillance. There is an increasing acceptance of the limitations of any pre-market clearance system—particularly for long-term implants—and this is focusing the minds of regulators on procedures for the follow-up of devices in service. Requirements for post-market surveillance of some kind is found in all three systems considered here although the precise form varies. The Global Harmonization Task Force is a forum in which discussion might be expected to lead to some convergence in the next few years (see Chapter 10).

- Exemptions for devices intended for clinical investigation are found in all three systems. Exemptions for custom devices are found in the European and US regulations. Submission requirements for permission to carry out clinical trials are similar.

The extent to which these features are found in recent regulatory developments in other countries, and which might therefore be regarded as universal, is explored in Chapter 5 and their place in a possible future global regulatory system is considered in Chapter 10.

Chapter 5

The current situation: regulatory developments in other countries

Several countries other than those which are the main focus of this book, i.e. the European Community, Japan and USA, are in process of introducing or revising medical device regulations. Several factors seem to be responsible: economic and technical developments have resulted in medical devices appearing in countries from which they were formerly absent; public anxiety has been aroused by the Bjork-Shiley heart valve (Lindblom *et al.* 1989) and breast implant (Snyder 1997) problems; the influence of the US and EC regulations is being felt in countries exporting to these regions; publicity about the Global Harmonization Task Force, etc.

Some awareness of the state of medical device regulation in the world at large, and the way in which developments are taking place, is essential for a full grasp of the subject. The state of these developments so far as is known at the end of 1st Quarter 2001 is summarized in this chapter. It should be noted that there is limited published information on many of the countries discussed in this chapter and, where sources are not acknowledged, the information is drawn from internal company reports of Quintiles—MTC.[1]

EEA and EFTA

The European Community has special links with other European countries which are members of the European Economic Area (EEA) or the European Free Trade Area (EFTA). The EEA Agreement (EC 92c) was signed on 2 May 1992 between the EC and the then members of EFTA (Austria, Finland, Iceland, Liechtenstein, Norway, Sweden and Switzerland). It extended the EC provisions for freedom of movement of goods, services, capital and labour, as embodied in the Single European Act (see Chapter 2), to these

---

[1] Quintiles Consulting—MTC, Action International House, Crabtree Office Village, Eversley Way, Egham, Surrey TW20 8RY, United Kingdom.

countries. In effect, this means that the relevant EC Directives are transposed in these countries, but they have not joined the EC Customs Union, neither are they covered by EC arrangements for VAT and excise duties. The Agreement is only concerned with the EC and does not extend to the EU's Common Foreign and Security Policy and Cooperation on Justice and Home Affairs.

The EEA was due to be established on 1 January 1993 but Switzerland voted in a national referendum against participation. This caused some rene-gotiation of the Agreement which was finalized in a protocol signed in March 1993 and the EEA came into effect on 1 January 1994. Since then, Austria, Finland and Sweden have joined the EC.

The three non-EC EEA countries, i.e. Iceland, Liechtenstein and Norway, have transposed the EC Directives into their national laws and Norway has established two Notified Bodies.

Switzerland, which is not party to the EEA Agreement, began a Mutual Recognition Agreement with the EC on 26 February 1999 and has transposed the MDD into national law. A new Federal Law on Therapeutic Products is in preparation and is expected to come into effect during 2001. This Law will establish a new Federal Institute for Therapeutic Products and will include provisions for the transposition of the IVDMDD into Swiss law.

Australia (and New Zealand)

Prior to 1989 there was no comprehensive national regulation of medical devices in Australia. In that year the Therapeutic Goods Act 1989 was passed (Australia 89) and the Therapeutic Goods Administration was set up to administer it. The Act covered medical products of all kinds, including drugs, devices and diagnostics. All medical devices must be registered or listed on the Australian Register of Therapeutic Goods. Nine categories of device are identified in Schedule 3 of the Regulations to the Act as registrable and are subject to pre-market evaluation for safety, quality and efficacy. Other (listed) devices require the submission of basic information such as compliance with quality system and/or product standards. All registrable devices and some listed devices are required to comply with a Code of Good Manufacturing Practice based on ISO 9000. Further details are given in Beech and Donovan (1995). The Act was amended in 1996 to improve definitions, to define offences against provisions of the Act, and to specify penalties for offences (Australia 96a).

The Australian system includes voluntary adverse incident reporting and 289 incidents were reported in the first six months of 2000 (Australia 00).

It was announced some years ago (Cable 1996) that the Therapeutic Goods Act is to be amended to match the EC Medical Devices Directive and that New Zealand is also to adopt a form of medical device regulation based on the MDD.

A Mutual Recognition Agreement (MRA) between Australia and the European Community was agreed during 1998 and announced in a Council Decision of 18 June 1998 (EC 98j). It came into force on 1 January 1999. Medical devices form one of the categories of products covered by this MRA and the Therapeutic Goods Act was amended on 7 July 1997 (Australia 97) to accommodate the MRA. This MRA does not imply that the regulations of the two parties must be recognized as equivalent but that the regulatory bodies in each party are able to carry out the procedures required by the other party, i.e. certain (to be) identified bodies in the EC will be able to carry out the procedures required for medical devices under the Therapeutic Goods Act, and the Therapeutic Goods Administration will be able to carry out the conformity assessment procedures of the EC Medical Devices Directive and the EC Directive on Active Implantable Medical Devices. A progress note from the European Commission (EC 00b) states that lists of EC conformity assessment bodies interested in participating in the MRA have been transmitted. An 18-month confidence building period is envisaged for ten high-risk products but it had not started in June 2000. An MRA on standards and conformity assessment between the European Free Trade Area (EFTA) and Australia was signed in April 1999 and came into effect on 1 July 2000.

Details of the operation of the MRA and of the proposed changes to the Therapeutic Goods Act are given in a recent issue of the Australian Therapeutic Devices Bulletin (Australia 98). Although this is not their prime purpose, the negotiations about the implementation of this Agreement will facilitate the revision of the Therapeutic Goods Act. An outline of the new regulatory system is given in the Australian Therapeutic Devices Bulletin No. 41, May 2000 (Australia 00) and a draft of the new regulations is expected to be available early in 2001.

A similar MRA between the EC and New Zealand was announced in a Council Decision of 18 June 1998 (EC 98k). This came into force on 1 January 1999 with no transitional period.

Canada

Medical devices have been regulated in Canada by the Medical Device Regulations (1978) (Canada 78) and the Food and Drugs Act 1985 (Canada 85). The Regulations were administered by the Medical Devices Bureau of the Department of Health. Devices were required to be safe and effective but pre-market review of evidence of safety and effectiveness was only required for a small number of devices listed in Part V of the Regulations. For all other devices, notification had to be made to the Bureau and evidence of safety and effectiveness kept available for inspection.

New regulations were published in 1998 (Canada 98) which came into effect progressively from 1 July 1998. These regulations have taken many elements from the EC Medical Devices Directive, and the main features are:

– All devices must comply with safety and effectiveness requirements which are essentially the same as the Essential Requirements of the MDD.
– Compliance with specified standards will be deemed to satisfy the essential requirements.
– All devices must be registered with the Medical Devices Program; the information to be provided with an application for registration depends on the class of the device.
– Devices are placed into four classes by rules virtually identical with those of the MDD.
– Manufacturers of Class II devices must be certified/registered by an appropriate organization as complying with ISO 13488 (incorporates ISO 9002); manufacturers of Class III and Class IV devices must comply with ISO 13485 (incorporates ISO 9001).
– Manufacturers of Class II, III and IV devices must have device licences; importers and distributors of medical devices must have establishment licences.
– There are requirements for labelling, distribution records and adverse incident reporting as well as specific conditions for the use of investigational and custom-made devices.

A Mutual Recognition Agreement between the EC and Canada was the subject of a Council Decision of 20 July 1998 (EC 98l). This is of the same form as that between the EC and Australia discussed above and came into force on 1 November 1998. By June 2000 a list of 14 EC conformity assessment bodies had been sent to Canada but no corresponding list had been sent to Europe. A transitional period to allow confidence building runs to 31 January 2001.

Eastern European Countries

Several Eastern European countries, formerly members of COMECON, are reviewing their medical device regulations which are, in general, based on drug procedures. Accession Partnerships have been concluded with Bulgaria, Czech Republic, Estonia, Hungary, Latvia, Lithuania, Poland, Romania, Slovakia and Slovenia (EC 98m). These are expected to lead to membership of the European Union and their adoption of the EC Medical Device Directives. The Accession Partnerships provide for an intermediate stage, the Protocol on European Conformity Assessment (PECA), under which mutual recognition of conformity assessment procedures will apply. The present position in those countries for which information is available is described below.

Bulgaria

The Medicinal Products Law covers both drugs and devices. The Law is administered by the National Drug Institute (NDI) which examines both the manufacturers (for quality assurance) and the products themselves. These examinations are confined to manufacturers in Bulgaria and result in manufacturing licences and market authorizations for the products. Clinical trials of medical devices may only be carried out with the permission of the NDI (Mircheva 1998). The NDI administers a registration scheme for implants and single-use devices. In accordance with the Accession Partnership (EC 98m), the Drug Law is being revised to transpose the EC medical device Directives. Possession of the CE marking gives marketing authorization 'significantly earlier than the stipulated 12 months approval process' (Popova 1998).

Czech Republic

The situation in the Czech Republic was similar to that in Bulgaria with medical devices regulated under a Medicinal Products Act (Czech 66) administered by the State Institute for Drug Control (SUKL) and amended in 1990, 1991 and 1992. Devices had to be registered with, and in some cases approved by, SUKL, the procedure being based on a declaration of conformity with the essential requirements of the EC medical device Directives or, in the case of IVDs, requirements published by the SUKL. Active (powered) devices could be required to undergo third-party testing. Decree No. 180/1998, published on 1 October 1998, began the process of transposing the EC Directives. This process was expected to be completed during 1999 (Michalicek 1998) but was delayed until 2000 with the publication of the Act for Medical Devices (Czech 00). The Competent Authority is now the Department of Pharmacy of the Ministry of Health. The role of SUKL is restricted to vigilance and the use and operation of devices in hospitals. A CZ mark is required; a CE Marking is helpful in obtaining the CZ mark but does not automatically obtain it. This situation will be overturned when the PECA agreement comes into effect.

Estonia

The current Estonian legislation covering medical devices consists of two laws: the Medicinal Products Act and the Procedure for Registration of Medicinal Products. These will be revised to match the MDD. The Estonian State Agency of Medicines is responsible for devices administered by the patients themselves (e.g. inhalers, contraceptives) and other medical devices are controlled by the Medical Technology Bureau of the Ministry of Social Affairs. There is no system of manufacturing licences (Mircheva 1998).

Hungary

Hungary had the largest manufacturing industry for medical devices in Eastern Europe and probably the most advanced regulatory system. Specific medical device regulations are contained in Decree No. 14/1990 (Hungary 90a) of the Ministry of Social Welfare. All medical devices, whether made in Hungary or imported, must be submitted to the National Institute of Hospital and Medical Engineering (ORKI) (Hungary 90b) before being allowed on to the market. ORKI will decide on the tests to be carried out on the device and may choose not to repeat tests carried out by recognized certification bodies. ORKI will ensure that a medical expert opinion is obtained after the technical tests have been satisfactorily concluded, since a 'clinical trial' is compulsory for all products marketed in Hungary (Bolvary 1996). Manufacturers are inspected for quality systems to ISO 9000. Adverse incidents must be reported to ORKI. Hungary issued in 1999 (Hungary 99) a ministerial decree which transposed the MDD and the AIMDD identically into Hungarian law and which came into force on 1 April 2000. Unfortunately the new legislation does not replace the earlier regulations but exists in parallel with them, which means that the CE Marking is acceptable only for Class I devices (not sterile or measuring). All other devices have to obtain the H mark following conformity assessment by ORKI. Eventually, mutual recognition of certificates of conformity between Hungary and the EC will apply under the Protocol on European Conformity Assessment envisaged by the Accession Partnership (EC 98m). Finally, the Directives will apply fully when Hungary becomes a member of the EC, expected in 2005 (Darday 1998).

Latvia

Marketing authorization for medical devices is given by the Centre for Health Statistics, Information and Medical Technology of the Ministry of Welfare. There is no system of manufacturing licences. Clinical investigations must be authorized by the Central Committee of Medical Ethics of Latvia. Latvia is preparing to bring in legislation aligned with the EC Medical Devices Directive (Mircheva 1998).

Lithuania

According to Mircheva (1998), Lithuania claims to have transposed the MDD into a new national regulation 'Medical Devices—Procedure of Approbation'.

Poland

Medical devices have been regulated in Poland since 1960 when a Commission for Medical Products Evaluation was established. In 1968 the Research

Center for Medical Technology was created to carry out testing. Following the political changes in the country, the regulations were revised by the Health Care Establishments Act 1991 (Poland 91a); the Pharmaceutical Products, Medical Materials, Pharmacies, Wholesaler Outlets and Pharmacy Inspection Act 1991 (Poland 91b); and the 1994 Directive on Registration of Drugs and Medical Materials (Poland 94). Under these regulations all medical devices are subjected to type testing and clinical trials at authorized laboratories and clinics. The results are reviewed by the Bureau of Drug and Medical Device Registration, part of the Polish Drug Institute, which issues approval certificates. Imported devices which are approved in another country can go through a simplified procedure. The authorized laboratories also carry out quality system checks against ISO 9000. Registration of laboratory diagnostics and electromedical equipment is conducted by the Center of Medical Techniques which is independent of the Drug Institute (Kielanowska 1996). The Polish Committee for Standardization, Measurement and Quality has two Technical Committees dealing with medical devices. The aim is to adopt standards from ISO, IEC, CEN and CENELEC and to introduce new legislation implementing the MDD (Katkiewicz 1994). Under an Agreement between Poland and the EC made in 1998 (EC 98n), mutual recognition of conformity assessment procedures will be introduced and Poland undertook to introduce EC Directives into its national legal system.

Russia

Medical devices must be licensed by the Public Health Ministry before being allowed on to the Russian market. Technical documentation is submitted to the All-Russian Research Institute for Medical Engineering (VNIIIMT) which decides on the technical and clinical tests to be carried out. Evidence from tests carried out elsewhere may be accepted but toxicology tests and clinical trials must be carried out in Russia. There is no provision for quality system checks. At the conclusion of the tests and clinical trials (which do not normally exceed two months), the documentation is sent for consideration by one of the specialized commissions of the Medical Device Committee which will decide if the device is to be licensed for manufacture or import. (Ischenko and Mihaylova 1994).

Slovac Republic

A new law governing both drugs and devices was introduced in 1998 (Slovac 98). The sections on medical devices are brief, as the requirements for approval will appear in a separate regulation understood to be in preparation. However, the law stipulates certain general rules: the definition of a medical device is very similar to that in the MDD; classification of medical devices again follows the MDD; only approved devices can be put into circulation and use; devices must be approved by the State Institute for Drug Control (SIDC) and approved devices will be registered in the List of Approved Medical Devices published by the SIDC. There are detailed requirements for clinical testing which may only be carried out under a licence from the SIDC. Although there is no official acceptance of EC certificates, CE marked products are approved faster (Martinec 1998).

Slovenia

A new Law on Medicinal Products and Medical Devices was issued on 17 December 1999. This replaces the Medicinal Products Act of 1996, under which medical devices were treated as a subset of medicinal products. The new law identifies devices as distinct from drugs and reflects the provisions of the EC medical device Directives. Implementing regulations are expected to come into force during 2000 and are likely to require devices in Classes IIa, IIb and III to be registered with the competent authority, the Agency for Medicinal Products and Medical Devices. A certificate of conformity with the MDD from a Notified Body is expected to be acceptable for registration.

Far Eastern Countries

Several Far Eastern countries are considering the introduction or amendment of their medical device regulations. A number of them, including China, Hong Kong, Korea, Singapore, Taiwan and Thailand, participate in an Asia–Pacific Harmonization Group (formerly the Asian Harmonization Working Party) initiated by the Health Industry Manufacturers Association (now the Advanced Medical Technology Association (AdvaMed)) of the USA and intended to prevent the development of disparate regulations in the region. The Group has made little progress so far, and requested the Global Harmonization Task Force in February 1998 for formal links to be established between the Group and the GHTF (Dick 1998) as the GHTF is likely to prove to be a source of guidance on harmonized aspects of regulatory systems (see Chapter 10). Current regulations in the region, where they exist, are briefly described below.

China

The regulatory situation in China has been subject to considerable changes in recent years. New 'Provisions' were issued by the State Pharmaceutical Administration of China (SPAC) in 1996 and came into force on 1 January 1997. These were followed by a period of uncertainty about where the responsibility lay for the control of medical devices. The registration of medical devices passed from the SPAC to the State Drug Administration, as described by Pemble (1998), and a new Medical Devices Act (China 00) was approved by the State Council on 1 January 2000, paving the way for Regulations for the Supervision and Administration of Medical Devices to come into force on 1 April 2000.

Under these Regulations, the definition of a medical device is essentially identical with the definition of a medical device in the European Medical Devices Directive (see Chapter 1). Devices are divided into three classes, according to risk, as in other administrations. This was the case under the preceding Provisions according to which they were listed in the Catalogue of Medical Device Products Classification (China 96) which may now be amended.

All medical devices must be registered for manufacture or import. The manufacturing registration of Class I devices is carried out by district drug regulatory authorities; the registration of Class II devices is carried out by provincial drug regulatory authorities; and that of Class III devices is carried out by the drug regulatory authority directly under the State Council. Clinical evaluation must be carried out for Class II and Class III devices before they are put into production. Domestic manufacturers must obtain a Medical Device Manufacturing Enterprise License. Imported devices must obtain an import product registration certificate from the drug regulatory authority under the State Council. This will be given on the basis of technical data and approvals given in the country of manufacture (there is no specific mention of FDA approvals or CE markings). Importers of Class II and Class II devices must obtain a Medical Device Distributing Enterprise License from the provincial or central government drug regulatory authorities.

Other provisions of the Regulations include quality systems for manufacturers, the accreditation of testing bodies and an adverse event reporting requirement.

Korea

Medical devices are regulated under the Pharmaceutical Affairs Law, introduced in 1954. This Law covers drugs, medical devices, cosmetics and sanitary aids, and comprises enforcement ordinances, regulations and notifications. New enforcement regulations for medical devices were published in September 1997, with a transitional period of two years to allow manufacturers to comply with them.

Under the new regulations medical devices are placed into three classes according to the Korean Classification for Medical Devices. Class I devices must be notified to the Korean Food and Drug Administration (KFDA) and must have a certificate of free sale from the country of origin. Class II and III devices must be approved by KFDA following a dossier examination and, if necessary, testing by the Korea Institute of Industrial Technology (KITECH). Clinical study data must be included for Class III devices which employ new technology, which involve a change of use for an already approved device or 'whose safety is continuously the most important factor'. In addition, all importers must obtain an import licence and must have a quality system checked by KITECH.

Approval requirements are simplified for devices which are substantially equivalent to approved devices or for which there are satisfactory test results from acceptable laboratories and/or clinical trials results from acceptable sources. Quality system checks are simplified for importers who can show that they meet ISO standards or the US or Japanese GMP.

The new regulations introduce tracking of devices to hospitals and include requirements for adverse event reporting and post-market surveillance (see Chapter 10).

Malaysia

At present (mid-2000) there is no regulation of medical devices in Malaysia, except for borderline products such as blood bags containing anti-coagulant, visco-elastic products, wound care/dressing materials, contact lens solutions and lubricants which require registration under the Control of Drugs and Cosmetic Regulations of June 1984. In response to the growing need to ensure the safety and effectiveness of medical devices, a proposal was presented to the Planning and Development Committee of the Ministry of Health in August 1996, and approved for implementation under the 7th Malaysian Plan (1996–2000).

A draft Medical Devices Act and Regulations were produced in 1997 after a study of the regulations in the EC, USA, Japan and Australia. It is understood that the draft Regulations address classification, essential requirements, conformity assessment, labelling and packaging, post-market surveillance and quality systems for design and manufacture. Despite the economic slowdown, the Ministry of Health is still aiming to bring in the new legislation during the 7th Malaysian Plan.

Taiwan

Medical devices are regulated under the Pharmaceutical Affairs Law. Devices which formerly required pre-market approval (PMA) were identified in the *List of Medical Devices Requiring Registration and Market Approval* and the *Guidelines for the Application of Imported Medical Devices for Registration and Market Approval*. Lists were also published of devices (such as most IVD reagents) which did not require pre-market approval. Any device not found on either list was the subject of an *ad hoc* decision by the Pharmaceutical Affairs Bureau (PAB) of the Department of Health.

The Department of Health is about to implement a GMP standard based on ISO 13485/8 (ISO 96a, 96b) and is developing new medical device regulations which were expected to be implemented during 1999 (Chung-Hwei 1998). However, this date was not achieved: a classification scheme, based on the US classification system, was announced on 2 May 2000 and the final rules for device registration were to be published by the end of June 2000. The registration system is expected to require GMP for Class I devices; GMP plus premarket registration for Class II devices; GMP plus premarket registration and submission of clinical data for Class III devices.

Thailand

Medical devices, including in vitro diagnostics, are regulated under the Medical Device Act BE2531 of 13 May 1988. This Act empowers the Minister of Health to issue ministerial notifications concerning medical devices. Examples of such notifications are Notice No. 6, which relates to medical devices prohibited from importation into Thailand, and Notice No. 19 relating to medical devices subject to pre-market approval.

Medical devices are placed into one of three classes according to a ministerial notification. For Class I devices a certificate of free sale in the country of origin is required. For Class II devices, some technical details must be submitted to the Food and Drug Administration. For Class III devices, such as condoms, hypodermic syringes, surgical examination gloves, HIV test kits, a licence is needed. This is granted on the basis of submitted technical information and compliance with Thai or international standards.

Thailand is currently in the process of developing new medical device regulations.

South America

In many of the South American countries, medical device regulations have existed for years within legislation for food, drugs and cosmetics. However, because of the difficulty of enforcing the regulations, medical devices were simply neglected (WHO 87). Although medical device regulations exist in Argentina, Brazil, Colombia, Ecuador, Peru, Uruguay and Venezuela, Larkin (1998) suggests that they are enforced only in Argentina and Brazil. An attempt is being made to change this situation within MERCOSUR (Mercado Commun del Sur), the Southern cone Common Market trade agreement.

MERCOSUR

The members of MERCOSUR are Argentina, Brazil, Paraguay and Uruguay. Chile has been accepted for membership and Bolivia is expected to join in the near future (Pitta 98). Government and industry representatives from the member countries have been working for several years on harmonized GMP and product registration requirements for medical devices. At this time (mid-2000) a GMP document, based on ISO 9001/2 has been published and approved by the four members (MER 95), together with guides on its implementation. A harmonized medical devices registration system (excluding in vitro diagnostics) has been approved by the four members and is being edited for publication.

The registration system is based on the examination of a standard dossier which includes a certificate of compliance with the GMP requirements and a certificate of free sale for imported products. There is a risk-based classification system with three classes. Class I devices are exempt from the GMP requirements and Class III devices must have clinical data. The system is modeled on the US requirements and approval by the FDA will ease the registration process. The regulatory agencies will be the Ministries of Health in the member countries and registration in any country will apply throughout MERCOSUR. The requirements become effective after each country publishes them in their official gazette; an effective date of the end of 1998 has been projected, but Pitta (1998) suggests that training of GMP inspectors will not be completed by that date.

Argentina

Medical devices are currently regulated by Resolution 255 (Argentina 94) which is administered by ANMAT (National Administration for Medicine, Food and Medical Technology, Ministry of Health and Social Action). Registration is given on the basis of a detailed technical dossier which must include a GMP compliance certificate and a certificate of free sale for imported devices. Devices approved by the FDA or bearing the CE marking are accepted. Devices manufactured in Canada, Germany, Japan, UK and USA are exempted from GMP inspection (Larkin 1998). The current system will be superseded by the MERCOSUR system in the near future.

Brazil

Medical device regulations in Brazil are based on a product registration system, administered by the Secretariat of Sanitary Inspection (SVS) of the Ministry of Health, introduced in 1993 (Brazil 93). Supporting regulations dated 1994, 1995, 1996 and 1997 deal with details such as classification, quality systems, registration forms and exemptions. Again, registration is given on the basis of a technical dossier, but somewhat less detailed than that for Argentina. A certificate of free sale is required for imported products.

Mexico

Although it is not strictly a South American country, Mexico is included here for convenience. All medical devices sold in Mexico must be registered with the Secretariat of Health (SALUD). Applications must contain technical information about the product including raw materials, any Mexican standards that are met and, for imported products, an export certificate or certificate of free sale and the identification of the Mexican distributor. Under the North American Free Trade Agreement, attempts are being made to harmonize Mexican requirements with those of the USA (HIMA 97).

South Africa

There are separate control systems in South Africa for active (electrically-powered) and non-active devices. The control of active devices is the more advanced. The Hazardous Substances Act (HSA) of 1973 (SA 73a) provides for the control of several groups of products, Group III being Electronic Products. Regulations Concerning the Control of Electronic Products (SA 73b), and Regulations Relating to the Sale of Group III Hazardous Substances (SA 89 and SA 91a) have defined a range of electromedical devices brought under the HSA. As the first medical products brought under the Act were radiation-emitting devices, the regulations are administered by the Radiation Control Directorate of the Department of National Health and Population Development.

Devices listed in the Regulations require a licence from the Radiation Control Directorate before they can be sold on the South African market. To obtain a licence, the manufacturer or importer must submit evidence of compliance with international standards (or equivalent) for the product and for quality control. The Directorate may conduct its own tests to verify the claims.

Non-active devices are controlled under the Medicines and Related Substances Control Amendment Act of 1991 (SA 91b) which was modified to define a medical device as a special category medicine. Registration of non-active devices is given by the Medicines Control Council (Muller 1994).

It was reported (Marcus 1996) that South Africa was preparing new medical device regulations, to include in vitro diagnostic products, which are likely to be based on the EC Directives. Such a regulation, establishing a South African Medicines and Medical Devices Regulatory Authority (SAMMDRA), was published in 1998 but subsequently withdrawn. It is expected that new or revised regulations will be promulgated during 2001.

Commentary

Activity world-wide on the revision of medical device regulations is at a level probably never reached before. Australia and the Eastern European countries are clearly adopting the European Medical Devices Directive;

South America and Mexico are aligned closely with the US system, and Canada and the Far Eastern countries (so far as their new systems are known) are taking over aspects of both.

A move from the regulation of medical devices under drug laws to specific medical device legislation is apparent. Features such as risk-based classification of devices and the requirement for manufacturers to have a quality system compliant with the ISO 9000 standards appear to be virtually universal. Compliance with product standards is a fairly general requirement but conformity assessment methods appear to vary considerably, depending on the expertise and resources available to the regulatory authority. Simple registration without any technical requirements is common for low-risk devices and, even for devices of higher risk, dossier examination is the only method possible in several countries.

It would be advisable for these countries to wait for the publication of documents now in progress within the Global Harmonization Task Force (see Chapter 10). These documents:

– Essential Principles of Safety and Performance for Medical Devices
– Role of Standards in the Assessment of Medical Devices
– Recommendations on Medical Device Classification
– Summary Technical File
– Guidelines for Regulatory Auditing of Quality Systems of Medical Device Manufacturers
– Adverse Event Reporting Guidelines for Decisions by Manufacturers and their Representatives

address most of the key elements of a regulatory system for medical devices and are based on experience and expertise not available in many of the countries discussed in this chapter. If these countries were to accept these documents and use them in the design of their own systems, a significant step would be taken towards a global regulatory system as discussed in Chapter 10.

Any future global system should be an evolution rather than a revolution and, accordingly, should incorporate elements clearly shown in all the regulations considered in this and the previous two chapters. These include risk-based classification of devices, dossier preparation and submission, and testing—both physically and clinically—of the more highly classified products. Other factors appear in the regulations but to varying degrees and in different forms. Four of these factors: quality systems, product standards and their use, the question of 'effectiveness', and post market controls, seem to the author to need further analysis and discussion in order to decide just what their place should be in medical device regulation. These four 'key factors' are discussed in the following four chapters.

Chapter 6

The place of quality systems

Quality systems is the first key factor to be examined in an attempt to assess the place that such systems should occupy in a modern medical device regulation.

Quality systems

Until well after the Second World War, most manufacturing industries tried to ensure that they sold only satisfactory products on the basis of inspecting finished products, identifying defective units and scrapping or reworking them. This defensive technique has generally been abandoned in favour of controlling the production process (Bergman and Klefsjo 1994).

The control techniques are based on statistical methods pioneered largely by Shewart (1933) and Dodge and Romig (1959). The idea of taking corrective action when defective products were found was the beginning of quality control.[1] The extension of this approach into a management system designed to ensure that products (and services) were always fully satisfactory constituted quality assurance[2] as pioneered by Deming (1982).

Requirements for the use of quality control were introduced into NATO purchasing contracts in the 1950s and subsequently several major purchasers and manufacturers developed and used their own quality control and quality assurance documents up to, and during, the 1970s and 1980s (Sanderson 1996).

The first publicly-available standard for quality systems was British Standard 5750, which was published in 1979 (UK 79). This was the forerunner

---

[1] Quality control is defined in ISO 8402 (ISO 94d) as: operational techniques and activities that are used to fulfil requirements for quality.

[2] Quality assurance is defined in ISO 8402 as: all the planned and systematic activities implemented within the quality system, and demonstrated as needed, to provide adequate confidence that an entity will fulfil requirements for quality.

of an international series of standards, ISO 9000, first published in 1987, and revised and re-issued in 1994 (ISO 94a). A noteworthy feature of the revision was the considerable expansion of the requirements for control of the design and development stages.

The ISO 9000 series describes three kinds of quality system:

- ISO 9001, Quality systems—model for quality assurance in design, development, production, installation and servicing (ISO 94a);
- ISO 9002, Quality systems—model for quality assurance in production, installation and servicing (ISO 94b);
- ISO 9003, Quality assurance—model for quality assurance in final inspection and test (ISO 94c).

The contents of these standards are shown in Table 6.1.

Table 6.1 Requirements of ISO 9001, 9002 and 9003 (1994) (Clause numbers)

Title	ISO 9001	ISO 9002	ISO 9003
Scope	1	1	1
Normative reference	2	2	2
Definitions	3	3	3
Quality system requirements	4	4	4
Management responsibility	4.1	4.1	4.1
Quality system	4.2	4.2	4.2
Contract review	4.3	4.3	4.3
Design control	4.4	–	–
Document and data control	4.5	4.5	4.5
Purchasing	4.6	4.6	–
Control of customer-supplied product	4.7	4.7	4.7
Product identification and traceability	4.8	4.8	4.8*
Process control	4.9	4.9	–
Inspection and testing	4.10	4.10	4.10*
Control of inspection, measuring and test equipment	4.11	4.11	4.11
Inspection and test status	4.12	4.12	4.12*
Control of non-conforming product	4.13	4.13	4.13*
Corrective and preventive action	4.14	4.14	4.14*
Handling, storage, packaging, preservation and delivery	4.15	4.15	4.15
Control of quality records	4.16	4.16	4.16*
Internal quality audits	4.17	4.17	4.17*
Training	4.18	4.18	4.18*
Servicing	4.19	4.19	–
Statistical techniques	4.20	4.20	4.20*

* Less stringent than ISO 9001 or ISO 9002

Other standards in the ISO 9000 series give guidance on the choice and implementation of a quality system.

Good manufacturing practice

The US Good Manufacturing Practice Regulation of 1978

The first use of the principles of quality assurance in medical device regulation was made by the Food and Drug Administration of the United States in its Good Manufacturing Practice Regulation published in 1978 (USA 78). The use of this Regulation as a post-marketing control is described in Chapter 4. The contents of the GMP Regulation were (numbers are the part numbers of the Code of Federal Regulations):

820.1	Scope
820.3	Definitions
820.5	Quality assurance program
820.20	Organization
820.25	Personnel
820.40	Buildings
820.46	Environmental control
820.56	Cleaning and sanitation
820.60	Equipment
820.61	Measurement equipment
820.80	Components
820.81	Critical devices, components
820.100	Manufacturing specifications and processes
820.101	Critical devices, manufacturing specifications and processes
820.115	Reprocessing of devices or components
820.116	Critical devices, reprocessing of devices or components
820.120	Device labeling
820.121	Critical devices, device labeling
820.130	Device packaging
820.150	Distribution
820.151	Critical devices, distribution records
820.152	Installation
820.160	Finished device inspection
820.161	Critical devices, finished device inspection
820.162	Failure investigation
820.180	General requirements (records)
820 181	Device master record
820.182	Critical devices, device master record
820.184	Device history record
820.185	Critical devices, device history record
820.198	Complaint files

There is considerable coverage of the content of ISO 9001 (see Table 6.1) but numerous omissions. Nevertheless, this was the first attempt by a regulatory authority to impose principles of quality control, if not full quality assurance, on medical device manufacturers and many manufacturers, both in the United States and overseas, had to make significant changes in their operations in order to satisfy the GMP Regulation and stay in business. The derivation of this GMP Regulation from Good Manufacturing Practice already applied to pharmaceuticals is shown by the omission of design requirements and by the emphasis given to the cleanliness of buildings and personnel (to avoid 'adulteration').

The UK Guides to Good Manufacturing Practice

At the same time as the United States Food and Drug Administration was developing its GMP, the UK Department of Health and Social Security (DHSS) was moving in the same direction. The Scientific and Technical Branch (STB), the organ of the DHSS with responsibility for the safety and satisfaction of medical devices, was becoming increasingly concerned about establishing the sterility of single-use, disposable sterile products, then coming into widespread use.

Such products are made by mass-production techniques and final product testing cannot be used satisfactorily for checking the sterility of individual products for several reasons including the destructive nature of the test, the time needed for the test and the unreliability of sampling (unless the sterilizing process is very well controlled). These considerations led the STB to consider the control of sterilizing and associated manufacturing processes as a means of assuring the sterility of the end product and, working in conjunction with the trade association MEDISPA (Medical Sterile Products Association), they issued in 1979 a document: *Guide to Good Manufacturing Practice for Sterile Single-use Medical Devices and Surgical Products*. This document was quickly revised with the additional assistance of other trade associations (Duncan 1986) and was re-issued as the *Guide to Good Manufacturing Practice for Sterile Medical Devices and Surgical Products* ('the Blue Guide') (UK 81).

This document was offered for adoption on a voluntary basis by members of MEDISPA and subsequently by other manufacturers selling sterile products to the National Health Service. Influenced by the success of this Guide and the impact of the US GMP Regulation, the DHSS moved rapidly to the publication of further GMP Guides covering most categories of medical devices. Although these Guides remained voluntary, considerable pressure was put on manufacturers to comply with them by the introduction of the Manufacturers' Registration Scheme (MRS) described in Chapter 2, under which NHS purchases were directed increasingly to manufacturers on the register.

The advice given to the NHS by the DHSS was that all medical devices should be bought under contracts specifying compliance of the devices with appropriate British Standards (UK 82a). As there was no mechanism for checking such compliance, the value of these contractual conditions was questionable and the MRS was a response aimed at identifying those manufacturers who operated under conditions that made it possible for them to claim legitimately that their products did comply with the Standards. By the late 1980s, it was very difficult for manufacturers who were not on the DHSS Register to sell devices to the NHS and new manufacturers, wishing to enter the UK market, generally found it necessary to be accepted on to the Manufacturers' Register in order to make sales. The Scheme was, therefore, acting in some ways as a pre-market approval process although without any legal force.

It is worth noting that the first of the DHSS GMP Guides was based on the World Health Organization Guide to GMP for pharmaceuticals, but all the later ones were based on the 1979 British Standard for quality systems BS 5750 (UK 79), the forerunner of the ISO 9000 series.

The Japanese Medical Device GMP

The Japanese Ministry of Health and Welfare published a GMP for medical devices, based on their pharmaceutical GMP, in 1987 (Japan 87). The contents of this document were:

Article	1	Purpose
	2	Definition
	3	Quality assurance organization
	4	Preparation of product standard code etc.
	5	Product standard code
	6	Quality assurance standard code
	7	Manufacturing operations
	8	Records
	9	Complaints
	10	Instruction and training

As can be seen from the contents, this was a rather brief document addressing very little beyond the manufacturing processes. The 'quality assurance standard code' did not correspond to a quality manual, but was a collection of work instructions. The 'product standard code' corresponded to the 'device master record' of the US GMP.[3]

Enforcement of the GMP was the responsibility of the prefectural authorities who did not carry out inspections overseas. Some 2700 pharmaceutical

---

[3] Device master record (DMR) means a compilation of records containing the procedures and specifications for a finished product (USA 78).

inspectors were employed in prefectural governments for the inspection and guidance of manufacturers, importers and distributors of medical devices, drugs and cosmetics (Hirai 1989). Compliance with the GMP in the case of imported devices was the responsibility of the importer (see Chapter 4).

From GMP to Quality Assurance

Developments in Europe

The major step in the use of quality systems as a regulatory tool for medical devices was taken in 1988 with the publication of the Commission Proposal for a Directive on Active Implantable Medical Devices (EC 88) which included compliance with quality system standards in the conformity assessment procedures (see Chapter 2). This was followed by the 1989 Council Resolution on the Global Approach (EC 89a) and the Council Decision of 13 December 1990 (EC 90a), which established quality systems as an essential component of conformity assessment procedures for all New Approach Directives and the publication of the Medical Devices Directive in 1993 (see Chapter 3). Community policy on the use of quality systems in New Approach Directives was further emphasized by the publication of Certif. 95/2 on The European Quality Assurance Standards EN ISO 9000 and EN 45000 in the Community's New Approach legislation (EC 95b).

The adoption of quality systems as a pre-market approval procedure was not easily achieved. Countries, such as France and Germany, which had a long tradition of type testing to establish the safety of products of all kinds, including medical devices, found it difficult to accept that a manufacturer could give the authorities and the public the same assurance of product safety, even when operating a quality system corresponding to ISO 9001. Much argument in Commission and Council committees, led by the UK and manufacturers (Doolan 1989; IAPM 1991), was needed before the principle was accepted (for certain products) but the third party assessment of a design dossier for the most serious products (which may be regarded as a corruption of the ISO 9001 system) had to be accepted and is found in Annex II, Clause 4, of the two medical device Directives and as supplementary requirements to Module H of the Council Decision.

Particular requirements for medical devices

The ISO 9000 series of standards is intended for application to the supply of products and services of all kinds. Its developers in ISO Technical Committee 176 have consistently maintained that it can be used in all situations without addition or further interpretation (ISO 97a). This view has not been maintained by all the specialist interests using the ISO 9000 series, including the

medical device sector, and application documents, supplementary requirements and/or guidance documents have been produced for several industrial
sectors.

The European Standards bodies CEN and CENELEC set up a Working
Group[4] to examine the applicability of the ISO 9000 series to medical devices
with a view to the designation of harmonized standards for the medical devices
Directives. The ISO 9000 series had been adopted identically as European
standards, designated EN 29000, in 1987. CEN/CENELEC found it necessary
to introduce additional requirements to reflect the requirements of the medical
devices directives, as well as the familiar US and UK GMPs, and published
EN 46001 and 46002 '*Application of EN 29001/2 to the manufacture of medical
devices*' in 1993. Following the publication of the second edition of ISO 9000
and the change in the CEN designation of these standards,[5] revised versions
were published in 1997 (CEN 97a, CEN 97b), and EN 46003, relating to
ISO 9003, was published in 1999 (CEN 99).

The requirements added to ISO 9001 are shown in Table 6.2.

Table 6.2 Supplementary requirements for medical devices in EN 46001

Sub-clause of EN ISO 9001	Supplementary requirements in EN 46001
4.2.1	Documentation of authority and responsibility
4.2	Compilation of Device Master Record
4.4.4	Inclusion of regulatory requirements in design input
4.4.7	Documentation of design verification
4.5.2	Retention of obsolete documents
4.6.3	Retention of purchasing documents
4.8	Procedures to ensure traceability
4.9	Personnel hygiene
	Environmental control in manufacture
	Cleanliness of product
	Maintenance of manufacturing equipment
	Installation and acceptance procedures
	Recording of information about special processes
4.10.5	Recording of inspectors of implants
4.13.2	Concessions and reworking
4.14.2	Handling of complaints
4.15.1	Special storage conditions
4.15.4	Control of packaging and labelling
4.15.6	Recording of distribution of implants
4.16	Retention of quality records
4.18	Training of personnel involved in special processes
4.20	Review of sampling methods

[4] CEN/CENELEC Coordinating Working Group on Quality Supplements.
[5] To EN ISO 9000.

CEN and CENELEC also issued guidance documents on quality systems for medical devices:

- EN 50103:1995 Guidance on the application of EN 29001 (now EN ISO 9001) and EN 29002 (now EN ISO 9002) for the active (including active implantable) medical device industry.
- EN 724:1994 Guidance on the application of EN 29001 (now EN ISO 9001) and EN 46001 and of EN 29002 (now EN ISO 9002) and EN 46002 for non-active medical devices.
- EN 928:1995 Guidance on the application of EN 29001 (now EN ISO 9001) and EN 46001 and of EN 29002 (now EN ISO 9002) and EN 46002 for in vitro diagnostic medical devices.

International Standards

The developments in Europe were followed up at the international level. ISO Technical Committee 210: Quality Management and Corresponding General Aspects for Medical Devices published two standards: ISO 13485 and ISO 13488 *Quality systems—Medical devices—Particular requirements for the application of ISO 9001 and ISO 9002* (ISO 96a, ISO 96b). These are almost identical with EN 46001 and EN 46002 and replaced the corresponding European standards at the end of 2000 (as it is a principle of CEN and CENELEC to adopt international standards wherever possible).

ISO/TC 210 has also produced a guidance document, based on the three European guidance standards and on a document produced by the Global Harmonization Task Force: *Guidance on Quality Systems for the Design and Manufacture of Medical Devices* (GHTF 94), recently published as ISO 14969: *Guidance on the application of ISO 13485 and ISO 13488* (ISO 99a).

Developments in the USA

The US Food and Drug Administration announced in 1990 (USA 90b) their intention to consider revision of the 1978 Good Manufacturing Practice Regulation. This decision was taken in the light of the widespread acceptance of ISO 9001, the Safe Medical Devices Act of 1990, which gave authority to the addition of design control requirements to the GMP, and of a study carried out by the Center for Devices and Radiological Health (CDRH) (USA 90c) which showed that 44% of the quality problems that resulted in the recall of a medical device were design-related. A rather equivocal discussion paper was issued by the FDA in 1993 (USA 93b); however, as the proposals for amendment appeared, they were generally welcomed on the grounds that 'Firstly, the harmonization of U.S. medical device GMP requirements with European and other international standards will help move the industry in the direction of a more global medical device quality standard, enabling U.S. manufacturers to sell their products all over the world without having to meet different quality and recordkeeping requirements. Secondly, it certainly cannot hurt the U.S. medical device industry to have better documentation and procedures for controlling design, purchasing, and servicing' (Kahan 1994).

After several years of discussion and amendment in the light of comments from the public and from the Global Harmonization Task Force, the revised version was published in the Federal Register of 7 October 1996 (USA 96b). The preamble to what is now called the Quality System Regulation states that 'the agency (FDA) believed that it would be beneficial to the public and the medical device industry for the Regulation to be consistent, to the extent possible, with the requirements for quality systems contained in applicable international standards, primarily, the International Organization for Standards (ISO) 9001:1994 . . . and the ISO Committee draft (CD) revision of ISO/CD 13485 . . . (now published as ISO 13485 (ISO 96a)).'

The content of the new Regulation is consistent with, but not identical to, ISO 9001/13485, and a comparison is shown below:

FDA regulation Section		ISO sub-clause	Comments
820.5	Quality system	4.2.1	FDA omit requirement for a quality manual (but see 820.186).
820.20	Management responsibility	4.1.1	
		4.1.2	FDA is less detailed than ISO.
		4.1.3	
		4.2.2	
		4.2.3	
820.22	Quality audit	4.17	
820.25	Personnel	4.1.2.2	FDA adds feedback of device defects.
		4.18	
820.30	Design controls	4.4	FDA adds detail on design validation; dates and signatures for approvals; sections on design transfer to production, and a Design History File, but omits specific requirement for clinical evaluation[6].
820.40	Document controls	4.5	FDA adds dates and signatures for approvals but omits master list and retention of obsolete documents.
820.50	Purchasing controls	4.6	FDA adds documentation of activities but omits retention of documents.
820.60	Identification	4.8	FDA omits identification of returned devices.

[6] As the requirement for clinical trials is in the medical device regulations, see Chapter 4.

FDA regulation Section		ISO sub-clause	Comments
820.65	Traceability	4.8	FDA omits distribution records (but see 820.160).
820.70	Production and process controls	4.9	FDA adds inspection of environment and maintenance; sections on buildings and adjustment, but omits installation and sterilization requirements.
820.72	Inspection, measuring and test equipment	4.11	FDA is less detailed than ISO.
820.75	Process validation	4.9	FDA is much more comprehensive than ISO.
820.80	Receiving, in-process and finished device acceptance	4.10	ISO is much more comprehensive than FDA.
820.86	Acceptance status	4.12	
820.90	Nonconforming product	4.13	
820.100	Corrective and preventive action	4.14	FDA adds recording requirements but omits details of customer complaints and advisory notices (but see 820.198).
820.120	Device labeling	4.15.4	FDA is much more comprehensive than ISO.
820.130	Device packaging	4.15.4	
820.140	Handling	4.15.2	
820.150	Storage	4.15.3	
820.160	Distribution	4.15.6 4.3	FDA adds purchase order review, shelf life control, and distribution records.
820.170	Installation	4.9	
820.180	Records	4.16	
820.181	Device master record	4.2.3	ISO does not use this term but has a similar requirement.
820.184	Device history record	4.16	ISO is less comprehensive than FDA and does not use this term.
820.186	Quality system record	4.2.1	Corresponds to a quality manual.
820.198	Complaint files	4.14.1	FDA is more detailed than ISO.
820.200	Servicing	4.19	FDA adds documentation and analysis.
820.250	Statistical techniques	4.20	
	Contract review	4.3	FDA does not have a specific paragraph but see 820.160.
	Control of customer-supplied product	4.7	FDA does not have a corresponding paragraph

The correlation between the international standards and the US Quality System Regulation is close enough for manufacturers to be able to implement a quality system that will satisfy both.

The new regulation came into force on 1 June 1997 with the exception of the design control requirements of Section 820.30, which came into force on 1 June 1998. The year's grace was to allow manufacturers time to implement these new requirements. The inclusion of design control requirements is likely to strengthen the FDA's desire to have all applications for PMA or 510(k) approval (see Chapter 4) supported by recent, satisfactory GMP inspections (Kahan 1992).

FDA guidance on the new Regulation was published by Trautman (1997); design control guidance was issued on the internet in March 1997 (http://www.fda.gov/cdrh/comp/designgd.html), and draft process control guidance was made available on the internet during 1998 but withdrawn during 1999. A new guidance manual on the inspection of manufacturers was published in February 2001 (USA 01b).

Developments in Japan

In December 1994 the Japanese Ministry of Health and Welfare published a revised Quality Assurance Standard for Medical Devices (Japan 94b) to replace the 1987 version. This Standard had been heavily amended from a 1993 draft in response to comments from the Global Harmonization Task Force to bring it very close to the international standards ISO 9001 and (the then draft) ISO 13485. A comparison by Maehara (1995) showed the content of the new Standard to be aligned paragraph-by-paragraph with the international standards. The new Standard was brought into effect on 1 January 1995.

Developments in Canada

In May 1998 the Canadian Department of Health published new Medical Devices Regulations (Canada 98). These Regulations are discussed in more detail in Chapter 5 but the relevant feature is the requirement for applications for medical device registration to include an attestation by an acceptable organization that the manufacturing organization has a quality system which complies with ISO 13488 in the case of Class II devices and with ISO 13485 in the case of Class III and Class IV devices.

The Regulations, with some exceptions, came into effect on 1 July 1998. One of the exceptions is the requirement for compliance with quality systems which came into effect on 1 August 2000.

Developments in Australia

As discussed in Chapter 5, it is proposed to change the regulations in Australia to bring them into conformity with the EC Medical Device Directives, including the quality system requirements.

Developments in South America

MERCOSUR, the common market formed by Argentina, Brazil, Paraguay and Uruguay, has adopted a Good Manufacturing Practice document based on ISO 9000. (MER 95).

Developments in China

China has adopted ISO 13485 and ISO 13488 as national standards and compliance with these standards satisfied the quality system requirements of the 1996 Provisions and will, presumably, satisfy the requirements of the new Regulations.

Quality assurance today

The growth of quality assurance

Since the publication of the ISO 9000 series in 1987 the standards have been adopted as national standards in some 90 countries. The certification of manufacturers for compliance with the standards has grown into an industry: at the end of 1999, 343,643 ISO 9000 certificates had been issued (ISO 00a). A large number of organizations now offer certification/registration against the ISO standards. Many purchasing organizations (including manufacturers choosing component suppliers and sub-contractors) demand the possession of an appropriate ISO 9000 certificate.

This growth in the use of, and reliance on, quality standards has, in the past few years, been inhibited to some degree by questions about the ability of these standards, which address *systems* to assure the satisfaction of *products* and by concerns about the competence of some of the certifying organizations and the consistency of operation of the large number of certifying bodies.

The first of these concerns appears to stem from a lack of understanding of the standards and fear that they may not be properly applied. ISO 9002 and ISO 9003 have no design content and can only be used to ensure consistent manufacture of a design shown to be satisfactory by other means. ISO 9001 is intended to control both design and consistent manufacture. Sub-clause 4.4 defines a system which, *if properly implemented*, cannot fail to ensure that the final design meets the 'design input'. The guidance on the application of ISO 9001, given in ISO 9000-2 (ISO 93) says that 'design inputs are typically in the form of

– product requirements specifications, and/or
– product description with specifications relating to configuration, composition, incorporated elements and other design features.'

It goes on to say 'Details agreed between the purchaser and supplier on how purchaser and regulatory requirements will be met should be included.' If this guidance is followed and the design input is properly compiled, then a satisfactory product will result. As Fairbairn (1995) says, 'QA works not just because it is based on sensible and logical principles, but also because everybody signs up to it. It has become a codified discipline which everybody recognizes and accepts. QA establishes certain maxims and reflects priorities for which there is a consensus. Product quality is assured by these generally accepted QA principles then being applied by sound management practice.'

Doubts remain, however, because of uncertainties about whether the auditing of systems includes (or should include) checking the comprehensiveness of the design input and whether, even if this is attempted, the quality system certification bodies have the product expertise to carry out such checking adequately. This issue, then, also becomes one of the competence and consistency of the auditing body. (There are, of course, those who believe that the whole ISO 9000 concept is a mistake (Seddon 2000).)

Certification bodies and accreditation

The question of the competence of certification bodies and the consistency between them has received a great deal of attention in the last few years. A series of ISO/IEC Guides on testing and certification has been published since 1982 but has only recently included guidance specifically relating to quality system certification. In Europe, the EN 45000 series, which corresponds to the ISO/IEC Guides, has included such a specific standard (CEN/CLC 89) since 1989:

EN 45012
General criteria for certification bodies operating quality system certification.

ISO/IEC recognized the need for such a standard and issued Guide 62: *General requirements for bodies operating assessment and certification/ registration of suppliers' quality systems* (ISO/IEC 96a) and this document will probably be adopted by CEN/CENELEC as a replacement for EN 45012.

Six new work items involving the complete revision of the EN 4500 series have recently been announced by the Technical Board of CEN. It is hoped and expected that this work will be done in conjunction with ISO/ CASCO[7] and that the result will be identical European standards and ISO/IEC Guides.

---

[7] ISO/CASCO is the ISO Committee on Conformity Assessment.

There are also standards on quality system auditing from ISO/TC 176 which have been adopted identically in Europe:

ISO 10011-1 (EN 30011-1)   Guidelines for quality system audit. Part 1: Audit;
ISO 10011-2 (EN 30011-2)   Guidelines for quality system audit. Part 2: Qualification criteria for quality system auditors.
ISO 10011-3 (EN 30011-3)   Guidelines for quality system audit. Part 3: Audit programme management.

These formed the basis of guidance documents on quality systems auditing of medical device manufacturers issued by the Global Harmonization Task Force (see Chapter 10).

Adherence to these standards would undoubtedly improve the whole auditing and certification process but widespread belief that certification bodies comply with and use these Guides/standards demands confidence in the accreditation process by which the certification bodies are recognized as competent.

Most developed countries have an accreditation body charged with examining the structure, technical competence and operation of certification bodies and accrediting those assessed as satisfactory. To improve consistency in the accreditation process, a number of activities are in train.

ISO/IEC Guide 61 (ISO/IEC 96b) has been published and adopted in Europe as EN 45010: *General requirements for assessment and accreditation of certification/registration bodies*. It has since been supplemented by ISO/IEC TR 17010:1998: *General requirements for bodies providing accreditation of inspection bodies*.

The International Accreditation Forum (IAF) has been formed, under the auspices of ISO and IEC, and has adopted the following Mission Statement:

'IAF exists to facilitate world trade by working to eliminate technical barriers to trade through the process of accredited certification which is recognized throughout the world. This process works on the basis of open, transparent and consistent application of internationally agreed standards, and is such that it commands public confidence in its results.'

The European counterpart of the IAF is the European Cooperation for Accreditation (EA) which was formed in 1997 by the merger of the European Accreditation of Certification and the European Cooperation for Accreditation of Laboratories. Membership is open to national accreditation bodies in the countries of the European Economic Area and the European Free Trade Association which comply with the relevant standards in the EN 45000 series.

The European Commission has also demonstrated its concern about the accreditation of Notified Bodies in its documents EC 97c and EC 98f—discussed in Chapter 3.

These moves will take some time to be fully effective but should provide a better basis for full confidence in the quality assurance system as a means of ensuring consistently satisfactory products.

Quality systems in regulation

Quality systems in European Community Directives

The 1989 Council Resolution on the Global Approach (EC 89a) first established quality systems based on EN 29000 (ISO 9000) as a key element in New Approach Directives. The important place of this series of standards was confirmed by the 1990 Council Decision (EC 90a) on the conformity assessment modules. By this Decision 'the Council recognized the use of quality systems as set out in the EN ISO 9000 series of European standards, as a means of demonstrating conformity of products for which the Directives set out the safety levels' (EC 95c).

A further indication of the Commission's thinking is given in Certif. 95/2 on the place of the EN ISO 9000 and EN 45000 series of standards in New Approach legislation (EC 95b). In this document two reasons are given for the use of quality systems:

- 'firstly in order to assure that products continuously meet the technical and quality specifications or in order to meet customer requirements,
- secondly in order to obtain those results at the lowest cost.'

It is clear, therefore, that the Commission and the Council regard quality assurance as a satisfactory (and preferable) way of meeting the legal requirements for safe products laid down in Directives. Module H of the Council Decision offers full quality assurance (i.e. ISO 9001) as a means of satisfying both design and production requirements but somewhat undermines this position by allowing optional study and approval of the design by a Notified Body. Although this provision was presumably added to placate authorities wedded to type approval, it weakens the argument for quality systems and, because of its usage for Class III medical devices and active implantable medical devices, may be an obstacle to mutual recognition agreements and the harmonization of regulations.

A survey carried out by the European Commission in 1998 (EC 99e) showed that 3647 medical device manufacturers had had their quality systems accepted under the MDD, 1498 of these were examined against Annex II of the Directive (Corresponds to ISO 9001).

Quality assurance in US regulation

The Federal Food, Drug and Cosmetic Act requires manufacturers of medical devices to comply with the Good Manufacturing Practice requirements but does not define any time at which such compliance must be shown. Originally, compliance was checked after the products had been approved for sale but, as noted above, the practice has developed of requiring satisfactory compliance before accepting PMA and 510(k)[8] submissions from manufacturers. Thus, GMP compliance has become a pre-approval requirement, but it plays no part in the product approval process.

This is a fundamental difference from the European approach, but a draft US position paper (USA 97b) on an EU/US Mutual Recognition Agreement (MRA) on medical devices states:

'The FDA will explore the possibility for certain devices of relying on the new design control requirements of the FDA quality system/GMP regulation, in lieu of 510(k) requirements. A possible outcome could be that conformance with design controls might be the basis for marketing authorization for designated devices and/or modifications to designated devices.'

This paragraph is not found in the final MRA (EC 98i, USA 98c) but gives a clear indication of likely changes in the US attitude to the use of quality systems in the conformity assessment of products. This view is supported by the emphasis given by the FDA to the publication of guidance to manufacturers on how to comply with the new GMP Regulation. The FDA issued guidance, based on a document prepared by the Global Harmonization Task Force, on videotape and published it in a guidebook by Trautman (1997). This was followed up in March 1997 by specific guidance on the design control requirements of the GMP Regulation made available on the FDA website at http://www.fda.gov/cdrh/comp/designgd.html.

Quality assurance in other administrations

Canada has introduced, and Australia is preparing to introduce, medical device regulations that incorporate compliance with quality systems into the product approval process in much the same way as the European Medical Devices Directive.

China, Japan and MERCOSUR require compliance with quality systems as a prerequisite of product approval but such compliance is not accepted as forming part of the product approval process.

Risk management

The introduction of a requirement for manufacturers to carry out a risk analysis was one of the innovations of the EC Medical Devices Directive.

---

[8] This is the pre-market notification process—see Chapter 4.

This has been seen as an important step in the evolution of medical device regulation and might have been identified as another 'key factor' but has come to be treated as part of the design control stage of a quality assurance system. The guidance on design control originally issued by the US FDA in 1997 and adopted by the Global Harmonization Task Force with little amendment later in the same year (see Chapter 10) includes risk management as a governing feature of the design process. The European standards body CEN produced a standard on risk analysis in 1997 (CEN 97c) and this was used by ISO/TC 210 as the basis of an ISO standard. This was further developed to address risk management and was adopted in 2000 as an international (ISO 00b) and a European standard, replacing the corresponding standards on risk analysis.

Future developments

A new version of the ISO 9000 series was published at the end of 2000 (ISO 00c, 00d, 00e). The new version departs from the established structure by combining the former ISO 9002 and 9003 with 9001 in the new ISO 9001:2000 and adopts a process-orientated quality management system. These new standards will be adopted in Europe as the EN ISO 9000:2000 series. The medical device community has regarded these changes with suspicion that they would make the regulatory use of the standards much more difficult (Riley 2000)—at one stage it was thought that the 1994 versions would have to be retained for regulatory purposes. To adapt these new standards to the design and manufacture of medical devices is a task which is already in train. ISO/TC 210 has produced preliminary drafts of new versions of ISO 13485 and 13488 which are intended to be used with ISO 9001:2000 to define quality systems for medical devices. CEN will adopt the ISO standards at the same time and will withdraw EN 46001 and 46002.

Commentary

Compliance by medical device manufacturers with quality assurance standards is required by all the major administrations world-wide. National GMP documents are being replaced by national standards or legal requirements based on either ISO 13485 (which includes ISO 9001) or ISO 13488 (which includes ISO 9002).

Since its introduction by the US in 1978, the role of GMP, which corresponds approximately to ISO 9002, has been to ensure that manufactured product consistently matches the sample or description that has been approved. That is a very valuable role which recognizes that type approval, or any other pre-market approval process, does not secure the safety of patients unless followed up by a system of controlling production—and industries of all kinds have adopted quality systems as the most effective and economical method of applying such control.

The New Approach Directives of the European Community have recognized the power of the design control requirements of ISO 9001, particularly the 1994 second edition which expanded these requirements, and have given this standard a place in the modules for assessing conformity with the requirements of the Directives. In the case of medical devices, ISO 9001 must be supplemented by the particular requirements of EN 46001. (In the near future, the European standards will be EN ISO 9001:2000 plus EN ISO 13485:2000 or EN ISO 13488:2000.)

The approach to conformity assessment adopted by the Community has not received total acceptance within the Community (Huriet 1996) or elsewhere (USA 96a). However, the doubts expressed relate as much to the competence and control of the Notified Bodies as to the effectiveness of quality systems. The moves in train within the Community (EC 97c) and internationally to establish reliable and consistent levels of competence of certification bodies described above should make it possible to focus on the design controls themselves. A comprehensive discussion of the place of quality systems in regulatory conformity assessment is given by Suppo (1997) who concludes that, with special reference to in vitro diagnostic medical devices, 'This route to conformity is at least equivalent to the alternative conformity assessment procedure of product testing, that is, type testing/ product approval in combination with batch testing.'

The Suppo view, which is shared by the author, is that the design control requirements of ISO 9001 and regulations based on this standard, if properly followed, ensure not only that the final design meets all applicable requirements (verification—Sub-clause 4.4.7 of the standard) but also that it satisfies the needs of the users of the medical device (validation—Sub-clause 4.4.8 of the standard). This subject is discussed further in Chapter 8.

All the evidence available at this time demonstrates that quality systems are accepted in all administrations with reasonably modern medical device regulations as an essential part of the regulatory regime. The quality system requirements invoked are either the ISO 9000 standards plus the ISO particular requirements for medical devices or are based on these standards. The requirement for inspection or certification of the quality system is becoming a condition for approval of the manufacturer's device but the use of quality systems as part of the conformity assessment procedure is still restricted to those countries accepting the Medical Devices Directive as a model. The use of authorized third parties to carry out the quality system inspections is similarly restricted in its application.

It should be noted that, although the EC Directives are the most forward-looking in terms of the use of quality systems in conformity assessment of products, compliance with quality system requirements is not mandatory in Europe. This is in contrast to other countries, such as USA and Japan, where such compliance is required by the regulations but is in addition to, rather than part of, the process of approval of products.

The European situation is a result of the principle enshrined in the 1990 Decision on the conformity assessment modules (EC 90a) that the Council should leave as wide a choice as possible to the manufacturers and should avoid restricting it to only product certification or to quality system certification. The growth in the use of quality systems since that time is likely to lead, in due course, to a more pronounced preference for quality systems as foreshadowed in CERTIF 95/2 (EC 95b) on the use of quality standards in Community legislation: 'The decision to leave the choice between quality system and product certification was directed towards allowing manufacturers, in particular SMEs, to operate under the Directives and to leave them the time to move over to the quality system if and when they were ready, and according to their own economic interests.'

As discussed in Chapter 4, signs are now being given that the USA is considering a gradual move to the use of both third parties and quality systems in conformity assessment. Such a move by the USA would influence other countries and it will be argued in a later chapter that these two elements of the MDD will prove to be important features of a future global medical device regulatory system.

Chapter 7

The use of product standards

Introduction

The second feature of recent developments in regulation to be identified as a 'key factor' is the use of compliance with voluntary product standards as a means—and, generally, a preferred means—of demonstrating compliance with the regulatory requirements.

The introduction of the 'New Approach' in Europe (see Chapter 2) gave a new prominence to the principle of reference to standards and has probably been the most important use of this principle. Several other countries are now adopting the same approach.

This Chapter traces the history of the use of standards, both informally and in legislation, explains product standards and how they are produced, and suggests that future medical device regulation is likely to be based on compliance with standards from the international standards organizations.

Background

A standard is defined in ISO/IEC Guide 2 (ISO/IEC 91) as:

> 'A standard is a document, established by consensus and approved by a recognized body, that provides, for common and repeated use, rules, guidelines or characteristics for activities or their results, aimed at the achievement of the optimum degree of order in a given context.'

This definition does not identify many of the features now recognized as characteristics generally found in standards from 'recognized bodies', and found in some alternative definitions, such as:

– standards are established by a consensus which must include *all* the interested parties
– the standardizing procedure is open and transparent to all interested parties
– standards are offered for public comment during their preparation
– published standards are subject to regular review (leading to confirmation, amendment or withdrawal).

These features reflect the fact that, although standards-making began (in Europe) by the industrial associations developing specifications for features (such as connectors) which served their common interests, they soon recognized that their status and value—and hence the benefits they brought to industry—were greatly enhanced if purchasers and users of their products took part in the process and if the process were carried out in an open and accessible way. Hence, all the major standards institutions now have similar rules of procedure by which it is open to anyone to propose a standards project and to take part in it. The standards programme is published and it is usually incumbent on the officials of the institution to make positive attempts to identify and involve all the possibly concerned interests. At some stage, every draft standard must be made available for public comment.

It is these features which distinguish a standard from any other kind of technical specification such as exists within any manufacturing enterprise, may form part of a manufacturer's promotional literature, and will normally be a key part of a purchasing contract.

'Recognized bodies' include:

– national standards bodies, such as the American National Standards Institute (ANSI). Many countries have two national bodies: one for electrical products and one for non-electrical products
– regional standards bodies, such as the European Committee for Standardization (CEN) and the European Committee for Electrotechnical Standardization (CENELEC)
– international standards bodies, such as the International Organization for Standardization (ISO) and the International Electrotechnical Commission (IEC).

Many of these organizations have been in existence for more than one hundred years but involvement with medical devices is much more recent. The first national standards for medical devices were produced in the 1950s, the first ISO medical device committee was established in 1951, and in 1972 the IEC Committee on Medical Electrical Equipment (TC 62) was set up. In Europe, the standards programme for medical devices began with the formation of CEN TC 55 (Dentistry) in 1971 and of CENELEC TC 62 (Medical Electrical Equipment) in 1968.

The development and content of a standard

The development process

The time and effort involved in producing a standard is considerable—it commonly takes five years to complete a national standard and even longer in the case of an international standard. It is an activity, therefore, which should not be embarked upon unless there is a good reason for doing so.

All standards bodies have a procedure for commencing a standards project which is aimed at ensuring that the work is justified and at establishing priorities. The procedure, such as that described in the ISO/IEC Directives (ISO/IEC 97), will seek to establish:

(a) The purpose of the proposed new work:

  – mutual understanding and communication
  – safety, health, protection of the environment
  – achievement of interchangeability or interface or compatibility provisions
  – performance, function, quality
  – economy of energy and raw material
  – variety control (rationalization)
  – consumer protection
  – other purposes

(b) The feasibility of the standard—this is closely linked with the scope of the proposed work and its timing. Standardization too early can restrict technical development and may simply not be supported by enough information and experience. Standardization too late may be costly for some producers who are then required to make significant changes in their products. These pitfalls can often be avoided by restricting the scope of the standard to just those features which can profitably be fixed, for instance a connector pattern or data format, while allowing freedom for technical development around these features.

(c) The support of interested parties—the work cannot be accomplished without the active participation of those involved in arriving at the desired consensus: the manufacturers, consumers and, possibly, regulatory authorities. The commitment of all these parties must be confirmed before work can start.

If sufficient support is received for the project to go ahead, the work will be allocated to a Technical Committee (TC); possibly an existing one can take on the work or a new one will be formed.

It is generally the case that the TC is the key element of the standards institution. A TC is normally formed to cover a substantial subject requiring a series of standards to be developed and maintained over a period of time.

Practice varies from body to body, but usually a TC is not formed just to write one standard. In most standards organizations a TC has full responsibility for the content of a standard, although this may be delegated to a Sub-Committee (SC).

Technical Committees have a long-term existence. Unless they are highly specialized, they tend to become management units which form Working Groups (WG) to write the individual standards. The WG is intended to have a temporary existence and be disbanded when the work is done.

The Technical Committee has the job of defining the standard to be written by the Working Group, approving the constitution of the WG, monitoring the progress of the work, deciding that a draft standard is ready for issue for public comment, reviewing the comments and deciding whether the draft is acceptable for publication (or for voting in the case of an international standard).

A Technical Committee is usually formed by invitations to the member bodies in the case of an international or regional organization and, in the case of a national organization, to bodies representing the interests concerned: trade associations; professional institutions; consumers' associations; Government; etc. The members are normally expected to represent the views of their nominating body. A Working Group will include members of the TC and individuals invited because of their expertise. Members of a WG are expected to work as technical experts and not to have any representational role.

Typical content

An arrangement often used for product standards (as described in the ISO/IEC Directives and with comments in parentheses by the author) is as follows:

Scope   (A crucial part of a standard. It must be clear to all interested parties when the standard applies and when it does not. In general, the broader the scope of the standard, the more impact it will have, but the feasibility and speed of producing a standard may be improved considerably by reducing the scope.)

Normative references
Definitions
Symbols and abbreviations
Requirements   (The requirements may be expressed in many ways, depending on the nature of the standard. In the past, many standards described precisely the way in which an object was to be made. It is now recognized that standards of this type restricted technical progress and often resulted

in technical barriers to trade. The preference now is to describe the attributes that the object must possess and leave the manufacturer full freedom in the manner in which they are achieved.)

Sampling
Test methods

(Standards written in this way must have a test for each requirement with clear pass/fail criteria. Again, it is desirable to allow the manufacturer freedom to demonstrate by any suitable means that the requirement is met but, in the event of a dispute, the test method contained in the standard is definitive. Standards describing the performance of a medical device, where this can be distinguished from its safety, should not have limiting values but should describe test methods and the way in which performance is to be declared so that the purchaser can make valid comparisons between competing products and thus make an informed choice.)

Classification and designation
Marking, labelling, packaging

(The method of identifying a product which is in compliance with the standard is normally included. It is particularly important for medical products to define the way in which the product and its proper use should be identified. The standard may also specify other information, such as installation or maintenance instructions, which should also be provided.)

Normative annexes.

Types of standard

Most standards describe characteristics of a particular product but there are so many products that the preparation of individual product standards is impossible and is also inefficient when many products share the same, or closely similar, characteristics. The standards bodies have, therefore, given their attention to the development, wherever possible, of standards of broad application. The terms 'horizontal', for standards of wide application, and 'vertical', for individual product standards, have been in common use, but the European standards bodies have defined the following hierarchy for standards for medical devices:

Level 1   generic standards covering fundamental requirements common to all, or a very wide range of, medical devices ('horizontal');

Level 2   group standards which deal with requirements applicable to a group of devices ('semi-horizontal');

Level 3   product-specific standards which give requirements particular to one device or a small family of devices ('vertical').

The international standards bodies are adopting the same approach. ISO/ IEC Guide 63 *Guidance on the development and inclusion of safety aspects in International Standards for medical devices* (ISO/IEC 99a) states:

'... to create a coherent approach to the treatment of safety in the preparation of standards. The use of a hierarchy of standards will ensure that each specialized standard is restricted to specific aspects and makes reference to standards of wider application for all other relevant aspects. Such a hierarchy is built on:

- **basic safety standard**, including fundamental concepts principles and requirements with regard to general safety aspects applicable to all kinds or a wide range of products, processes and services. (Basic safety standards are sometimes referred to as horizontal standards);
- **group safety standard**, including safety aspects applicable to several or a family of similar products, processes or services dealt with by two or more technical committees or subcommittees, making reference as far as possible to basic safety standards;
- **product safety standard**, including all necessary safety aspects of a specific, or a family of, product(s), process(es), or service(s) within the scope of a single technical committee or subcommittee, making reference, as far as possible, to basic safety standards and group safety standards.'

Examples of basic safety (Level 1) standards are:

Biological evaluation of medical devices, Part 1: Guidance on selection of tests (ISO 10993-1).
Medical electrical equipment—Part 1: General requirements for safety (IEC 601-1).
Symbols to be used with medical device labels, labelling and information to be supplied (ISO 15223).

Examples of group safety (Level 2) standards are:

Non-active surgical implants—general requirements (ISO 14630).
Packaging for terminally-sterilized medical devices (ISO 11607).
Medical electrical equipment—Part 1: general requirements—3. Collateral standard: radiation protection (IEC 601-1-3).

Examples of product safety (Level 3) standards are:

Implants for surgery; metal bone screws with asymmetrical thread and spherical under-surface; mechanical requirements and test methods (ISO 6475).

Pulse oximeters for medical use; requirements (ISO 9919).
Particular requirements for the safety of medical electron accelerators in
the range 1 MeV to 50 MeV (IEC 60601-2-1).

National standards

The national standards bodies played an important role in expressing in a technical form the requirements for a limited range of medical devices until the late
1980s. The growth of the international standards movement during the 1970s
and 1980s began a process of changing the role of national standards bodies to
one of commenting on, and voting on, draft international and regional standards with a view to national adoption. This process was greatly stimulated
by the introduction of the 'New Approach' medical devices Directives (see
Chapter 3) and the Commission mandates for the development of standards
suitable for use in the regulatory system by being designated as 'harmonized'.
Today the major importance of national standards bodies derives from their
membership of the international standards bodies and, in the case of the
European countries, of CEN and CENELEC.

Nevertheless, it is useful to appreciate the part played by national
standards in some of the larger countries.

Germany

There are two standards bodies in Germany: for non-electrical products,
DIN, Deutches Institut fur Normung e.V. (German Institute for Standardization), and for electrical products, VDE, Verband Deutscher Elektrotechni-
ker (German Association of Electrical Engineers).

DIN was founded in 1917 and is recognized as being appreciably the
largest standards body in Europe. More than 25,000 published standards
are listed, 1000–2000 of these being in the healthcare field. The traditional
difference between DIN and VDE has diminished in recent years and most
electrical standards are now published as DIN/VDE standards. DIN is the
member body of ISO and CEN. The German Electrotechnical Committee,
which is the member body of IEC and CENELEC, responds to both DIN
and VDE.

United Kingdom

The British Standards Institution claims to have been the first national standards body in the world and has a catalogue of more than 10,000 standards,
some 300 of which apply to medical devices. It is one of the few standards
bodies which deals with both electrical and non-electrical products and is,
consequently, the member body of ISO, IEC, CEN and CENELEC.

British Standards for medical devices are produced under the overall management of the Healthcare Standards Committee which oversees the work of some 70 Technical Committees concerned with particular categories of product.

France

The two standards organizations in France are: for non-electrical products, AFNOR, Association Française de Normalization (French Standardization Association), and for electrical products, UTE, Union Technique de l'Electricité (Technical Union of Electricity).

AFNOR and UTE have a portfolio of more than 13,000 standards. Between 200 and 300 cover medical devices. AFNOR is the French member of ISO and CEN, and UTE is the member of IEC and CENELEC.

United States

In the United States, the organization of standards production and adoption has taken a different path from that in Europe. The first standards to be widely adopted nationally were developed by national professional bodies which took care of the needs of their particular technical sector. Examples of such bodies are ASTM (American Society for Testing Materials) and AAMI (Association for the Advancement of Medical Instrumentation).

Several bodies of this kind became well established in the US (and many of their standards came into use world-wide) but found it difficult to establish sound working relations with the international standards organizations as their importance grew in the post-Second World War period. ANSI (the American National Standards Institute) was formed in 1918 to become the focal point of US standards-making and the US member of ISO and IEC. ANSI is a private sector, not-for-profit organization, which, in contrast to the European model, does not have a monopoly of standards-writing committees but delegates most standards preparation to the professional bodies with established committees. Standards are then published with a designation which indicates the organizations which participated in the production of the standard, e.g. ANSI/AAMI/ISO 11135:1994, *Medical devices—Validation and routine control of ethylene oxide sterilization*. At the end of 1999, there were 14,650 published American National Standards.

Japan

Standards play an important role in Japan. The Japan Industrial Standards Committee (JISC) was established in 1953 under the Industrial Standardization Law of 1949. The JISC is the national member of both ISO and IEC.

Standards-making in Japan is dominated by industrial interests and the national standards are known as Japanese Industrial Standards (JIS).

According to Naito (1996), the total number of JIS is more than 8000. 315 of these are medical device standards, produced by 36 Technical Committees. About 30% of JIS for medical devices are aligned with international standards and the 8th Five Year Standardization Plan 1996–2000 includes a programme to increase this proportion.

International Standards

The two major international standards bodies are the International Organization for Standardization (ISO) and the International Electrotechnical Commission (IEC). ISO deals with non-electrical products; IEC deals with electrical and electronic products.

Cooperation between ISO and IEC has improved markedly in recent years. The two bodies have agreed joint ISO/IEC Directives which came into force on 1 February 1990. There is a Joint Presidents' Group on Policy and Organization and, in the field of medical devices, there is a Joint ISO/IEC Technical Advisory Group (JTAG) 'Healthcare Technology' which met for the first time in March 1990. The terms of reference of this JTAG are:

'To advise the IEC Committee of Action and the ISO Technical Board on matters of cross-sectoral coordination, coherent planning and needs for new work by means of:

(a) surveying and coordinating the programme of work of the IEC TCs and ISO TCs involved in the standardization of healthcare technology with a view to identifying risks of duplication and of unnecessary diversification
(b) proposing ways and measures for a rational distribution of work, including the development of joint standards when appropriate
(c) reviewing and assessment of priorities on the basis of strategic policy statements developed by the TCs
(d) overseeing questions of interface with other international organizations.'

This JTAG has largely succeeded in improving cooperation between medical device committees in ISO and IEC and remains an important liaison mechanism.

ISO

The International Organization for Standardization was formed in 1947 and has now members from 124 countries. 12,500 standards had been published at the end of 1999, approaching 500 of these in the health-care sector (ISO 00a). All national bodies have the right to participate in Technical

Committees and Sub-Committees and choose to participate actively (as 'P-members') or to follow the work as observers ('O-members'). Final Draft International Standards are circulated to all members for voting, and P-members of the originating TC are obliged to vote. A standard is approved if: (a) a two-thirds majority of the votes cast by the P-members of the Technical Committee or Sub-Committee are in favour, and (b) not more than one-quarter of the total number of votes cast are negative.

All votes are of equal weight. The adoption of an international standard as a national standard in any country is decided by the national standards bodies individually.

Healthcare activities are carried out in the following Technical Committees:

TC 76	Transfusion, infusion and injection equipment for medical use
TC 84	Syringes for medical use and needles for injections
TC 106	Dentistry
TC 121	Anaesthetic and respiratory equipment
TC 150	Implants for surgery
TC 157	Mechanical contraceptives
TC 168	Prosthetics and orthotics
TC 170	Surgical instruments
TC 173	Technical systems and aids for disabled or handicapped people
TC 194	Biological evaluation of medical and dental materials and devices
TC 198	Sterilization
TC 210	Quality management and corresponding general aspects for medical devices
TC 212	In vitro diagnostic systems.

Other ISO committees whose activities have some bearing on healthcare are:

TC 42	Photography (deals with radiographic films and screens)
TC 45	Rubber and rubber products
TC 58	Gas cylinders
TC 85	Nuclear energy (deals with protective enclosures)
TC 94	Personal safety—protective clothing and equipment
TC 172	Optics and optical instruments
TC 176	Quality management and quality assurance (responsible for the ISO 9000 series).

IEC

The International Electrotechnical Commission was established in 1906. Today, the national electrotechnical committees of 42 countries are members and, in the case of the European countries, they are also the members of

CENELEC. There are now 93 Technical Committees and a publication list of more than 5000 standards. The rules for participation in the standards work, and for the approval and adoption of an IEC Standard, are the same as those applying in ISO and described above.

Medical device activity is concentrated in Technical Committee 62 'Electrical equipment in medical practice', which was formed in 1968, and its four Sub-Committees. Each of these Sub-Committees has set up Working Groups, some of which have finished their tasks and have been disbanded. The four Sub-Committees are:

SC 62A   Common aspects of electrical equipment in medical practice

SC 62B   X-ray equipment operating up to 400 kV and accessories

SC 62C   High energy radiation equipment and equipment for nuclear medicine

SC 62D   Electromedical equipment.

TC 62 is responsible for IEC 601 '*Medical electrical equipment: requirements for safety*'. This has been probably the most important single medical device standard—not only because of its wide application but because it was the first example of a structure now being widely applied to medical device standards: a standard containing generally-applicable requirements (a 'horizontal' standard) supplemented by a series (now more than 50) of generally short standards containing the special requirements applicable to particular devices ('vertical' standards).

Other IEC TCs and SCs which have some bearing on medical devices are:

TC 64   Electrical installations of buildings

SC 65A   Industrial-process measurement and control and laboratory equipment

SC 66E   Safety of measuring, control and laboratory equipment

TC 76   Laser equipment

TC 77   Electromagnetic compatibility between electrical equipment including networks

TC 87   Ultrasonics

TC CISPR   International special committee on radio interference—Sub-Committee B: Interference from industrial, scientific and medical radio-frequency apparatus.

CEN and CENELEC

CEN (The European Committee for Standardization) and CENELEC (The European Committee for Electrotechnical Standardization) are non-profit-making institutions of a scientific and technical nature, registered in

Belgium.[1] Their members are the national standards organizations of the EC and EFTA countries, although affiliation, and eventual membership, is being extended to some East European countries.

The two institutions moved into common premises in 1975 and, together, they constitute the Joint European Standards Institution. The Joint Institution is governed by the CEN/CENELEC Agreement of 1982. The Joint Institution has produced Internal Regulations under which both bodies operate (CEN/CLC 94). The formal agreements with the European Commission and with the EFTA Secretariat are made by the Joint European Standards Institution.

CEN and CENELEC coordinate their work programmes with those of ISO and IEC (respectively) via the 'Vienna Agreement' between ISO and CEN of 1991 (CEN 91) and the IEC/CENELEC Co-operation Agreement of 1989 (CENELEC 89). Besides exchanges of information, these Agreements provide for new work proposals accepted in CEN and CENELEC to be offered to the corresponding international body for the development of an international standard, and for the resulting draft standards to be submitted for parallel voting in the European and international bodies. Most of the current work on medical device standards is being carried out under this arrangement.

Adopted standards must be implemented in their entirety as national standards, regardless of the way in which the national member voted, and any conflicting standards must be withdrawn. It is this feature, together with the 'standstill' agreement, which forbids members to undertake the development of national standards if corresponding work is going on in CEN/CENELEC, which makes the position of the European standards bodies much more powerful than that of the international organizations and emphasizes the importance of standards in Europe.

Medical device standards in CEN and CENELEC

In anticipation of requests from the European Commission, CEN set up a Healthcare Task Force in 1989 to take over the supervision of four existing Technical Committees (55, 102, 140, 170) and to plan the expansion of the work to meet the needs of the Directives. The Healthcare Task Force was superseded by a subsidiary of the Technical Board, and later by the CEN Healthcare Forum (CHeF), and the programme expanded to involve 17 TCs.

A medical device activity has existed in CENELEC since 1975. TC 62: Electrical Equipment in Medical Practice corresponds closely to IEC TC 62 and most of its work has been limited to the adoption of standards originating in IEC but the Directives programme has resulted in the creation of several Working Groups to develop specific European standards.

---

[1] A third body, ETSI (European Telecommunications Standards Institute), is responsible for standardization in the field of telecommunications.

The European Commission issued standardization mandates to CEN and CENELEC in 1990, 1991, 1993 and 1996. A mandate was issued to the European Pharmacopoeia in 1995 for a limited number of EP Monographs which will be given the status of harmonized standards. The current programmes have been developed in response to these mandates. A mandate issued in 1999 (EC 99a) expands the range of standards required and calls for the replacement and updating of some of the earlier standards.

For a more detailed account of CEN, CENELEC and the European health care standards programme, see Moore (1994).

Standards in regulations

In several European countries, compliance with standards has, for many years, been a way of gaining customer confidence in manufactured products. Proprietary marks of compliance with standards, such as the BSI 'kite' mark and the AFNOR 'NF' mark, have become important sales aids.

However, the principle that compliance with identified standards should demonstrate compliance with regulatory requirements was adopted much more slowly. The German Law on the Safety of Technical Products (Germany 68) may have been the first to use this principle. In 1982 the UK Government and the BSI signed a Memorandum of Understanding (UK 82b) that committed the BSI to the development of standards that would be suitable for reference in regulations and the Government to the use of British Standards, rather than its own technical specifications, in future legislation. This MoU was replaced by a revised version in 1995.

In terms of medical devices, the Danish Heavy Current Regulation of 1982 (Denmark 82), the German MedGV of 1985 (Germany 85) and the French homologation scheme (France 87 and France 90) were the earliest examples of reference to standards. The US Medical Device Amendments of 1976 (USA 76a) identified compliance with standards as an appropriate requirement for Class II devices but standards as specified in the Amendments were never produced by the FDA. The FDA Modernization Act of 1997 (USA 97a) introduced a provision for compliance with 'recognized' voluntary standards to contribute to the approval process. The Japanese Pharmaceutical Affairs Law has, for many years, accepted compliance with Japanese Industrial Standards as compliance with the regulations for certain low-risk products.

Standards in EC Directives

As discussed in Chapter 2, the EC programme for completion of the Internal Market (EC 85c and EC 86) was based on the New Approach to Technical Harmonization and Standards (EC 85b) of which the key principle was that compliance with 'harmonized standards' was deemed to satisfy the essential requirements of the Directives.

The development, therefore, of a corpus of harmonized standards addressing the essential requirements was crucial to the success of these Directives. The issuing of mandates and the expansion of the activities of CEN and CENELEC were priority tasks for the European Commission which retained a close supervision of the standards bodies.

In 1991, the Commission published a Green Paper on the development of European standardization (EC 91b). This paper gave rise to an extensive debate, the results of which were summarized in the Communication from the Commission on standardization in the European economy (EC 91c). On the basis of this Communication, the Council adopted, on 18 June 1992, a Resolution (EC 92d) in which it emphasized the strategic importance of standardization, confirmed a series of principles underlying European standardization, encouraged the use of European standards as an instrument of economic and industrial integration, and advocated a wider use of European standards in Community policy.

These initiatives stimulated CEN and CENELEC to examine their procedures to find ways of making the standards-writing process more efficient, and encouraged the Commission to review the issuing of standardization mandates. Today, about one third of the European standardization activities is covered by mandates from the Commission (EC 95d, paragraph 1.6) and the 'Monti Report' (EC 97d, Section 2.2.3.4) states that 'On the whole, the task of preparing European standards required for the effective functioning of the single market is well in hand . . . European standardization has registered a dramatic expansion in its activity (5000 European standards produced and 16,000 further work-items are foreseen).' Monti describes this as a 20-fold increase in annual output over 10 years.

In a Communication to the Council and Parliament in 1995 (EC 95d) the Commission reviews the place of standardization in Community policy. It discusses the merits of the use of voluntary standards as an alternative to legislation, states that the use of standardization should be encouraged, and that, 'where appropriate, the principle of referring to European standards in Union legislation should be used.'

Standards in the Medical Device Directives

Standards required for the efficient working of the Active Implantable Medical Devices Directive and the Medical Devices Directive have been developed under the mandates described earlier. Lists of standards identified as 'harmonized standards' under one or both of these Directives have been published every year from 1994, reaching a total of 162 at the end of 2000.

The status of these standards must be borne in mind. Compliance with applicable harmonized standards or parts of harmonized standards is deemed to satisfy the legal requirement of compliance with the Essential Requirements of the Directives. Nevertheless, it is not mandatory to comply with harmonized standards. Despite their special status, the standards remain voluntary and it is permissible to show compliance with the Essential Requirements by alternative methods.

Standards in non-European regulations

Japan

Japanese Industrial Standards are given a special status in respect of certain 'low risk' devices. Article 18 of the Enforcement Regulations (Japan 95b) states that devices listed in Table 1 (of the Enforcement Regulations) are exempted from the general requirement for the approval of medical devices. Table 1 includes 126 devices which are required to comply with JIS standards.

USA

The Medical Device Amendments of 1976 (USA 76a) provided that the requirements for Class II devices included compliance with performance standards. No performance standards have been promulgated by the Food and Drug Administration (FDA) and this provision has not been implemented. In 1994 the FDA published a draft policy on standards (USA 94) in which it outlined the conditions under which the agency would participate in domestic and (in particular) international standards activities, and the possible ways in which the agency might use the resultant standards. These included: serving as a basis for mandatory standards; incorporation into guidance documents, and incorporation into compliance policy guides.

In fact, many standards from a variety of sources have been used in FDA guidance documents. Since the early 1990s these have increasingly been international standards—the first of them probably was IEC 601-1 (IEC 77), the dominant world standard on the safety of medical electrical equipment. Compliance with this standard was demanded by several States and cities, even for devices which had received FDA approval.

A progress report on the development of the FDA's thinking about the use of standards was given by Marlowe (1997) early in 1997. He explained that a significant group of people in the Agency considered that a substantial part of the work involved in a 510(k) submission (see Chapter 4) could be addressed by conformance with a standard. This report was followed by the enactment of the Food and Drug Administration Modernization Act of 1997 (USA 97a) in November 1997. Among the provisions of this Act, which is discussed more fully in Chapter 4, was one authorising the FDA to recognize all or part of a national or international medical device standard. The Agency will have to accept a declaration of conformity to such a standard submitted by a manufacturer in support of a 510(k) pre-market notification. Guidance on the recognition and use of consensus standards was issued in February 1998 and announced in the Federal Register (USA 98a). At the end of 2000, 501 standards had been 'recognized' by the FDA.

Other countries

Australia has announced its intention to amend the Therapeutic Goods Act to match the EC Medical Device Directive. This will give harmonized standards the same 'deemed to satisfy' status as they have in Europe.

Canada published new Medical Device Regulations (Canada 98) which came into force during 1998. The Regulations are similar in several respects to the European Directives. For Class III and Class IV devices (corresponding to Classes IIb and III in Europe) applications for registration must include a list of standards used in the design and manufacture of the devices, and 'recognized' standards will have 'deemed to satisfy' status.

Article 2 of the World Trade Organization's Agreement on Technical Barriers to Trade (WTO 95) accords a priority role to international standards in the elimination of technical barriers to trade. It requires members in principle to use international standards or the relevant parts of them as a basis for their technical regulations, lays down that compliance with the relevant international standards creates a *prima facie* presumption of conformity to the Agreement, exempts draft technical regulations conforming to international standards from the obligation to notify them in advance to other members, and urges members to participate in the preparation of international standards with a view to harmonizing technical regulations.

The communiqué from the Sevilla meeting of the TransAtlantic Business Dialogue (TABD 95) states (paragraph 1.2) 'The EU and US should aim to develop and adopt common and open standards wherever possible based on international product standards such as those of the ISO and IEC' and in paragraph 1.5 that 'Technical regulations set by government should rely on functional standards.'

Discussion

The growth in the use of the principle of reference to standards in legislation from its introduction in the 1960s to its widespread use in the EC Directives and beyond is clearly shown in this chapter. The recent developments in the US, Australia and Canada, coupled with the increase in medical device standards work in the international organizations, suggest that compliance with international standards will soon be widely accepted as the preferred regulatory approach.

The principle that compliance with product standards is an appropriate means of demonstrating that medical devices are safe and suitable for use is a key feature discriminating medical devices from pharmaceuticals. This results from the development of product standards away from descriptions of the products themselves (e.g. materials, dimensions, manufacturing methods) towards descriptions of the behaviour of the products, particularly with respect to their safety (e.g. maximum electrical leakage current, minimum breakage strength, maximum surface temperature). The ISO/IEC Directives, Part 3 (ISO/IEC 97), requires that all relevant characteristics are covered, either explicitly or by reference, in the standard, that limiting values are stated, and that the test method for each characteristic is included or given by reference. Further guidance on the inclusion of specifically safety aspects in product standards is given in ISO/IEC Guide 51 (ISO/IEC 99b).

This approach can be used because the possible hazards[2] associated with medical devices, listed in ISO 14971, *Medical devices—Application of risk management to medical devices* (ISO 00b), arise mostly from characteristics which can be measured objectively in the laboratory. These possible hazards are:

— energy hazards
— biological hazards
— environmental hazards
— hazards related to the use of the device
— hazards arising from functional failure, maintenance and ageing.

'Safe', i.e. permissible, levels of many of these hazards have been determined by experiments on animals and humans. For example, IEC 479-1: *Effects of current on human beings and livestock* (IEC 94), includes a bibliography of work done to determine acceptable levels of electric current which might pass through the human body, and publications of the International Commission on Radiation Protection (e.g. ICRP 77) record work done to establish acceptable levels of exposure to ionizing radiation. Once such levels have been established, the safety of the medical device can be determined in the laboratory without tests involving patients. Only where devices are genuinely new, with the possibility of unknown hazards, is it necessary to carry out controlled trials on patients.

Biological hazards are the most difficult to address by means of standards. Nevertheless, this has been done. The ISO 10993 (ISO 94e) series of standards on the biological evaluation of medical devices gives guidance on the selection of materials and specifies the tests to be done according to the intended mode of use of the device. These tests are not as

---

[2] A hazard is defined in ISO/IEC Guide 51 (ISO/IEC 99b) as a potential source of harm, where harm is physical injury and/or damage to health or property.

straightforward as physical tests, and generally involve the subjective assessment of damage to animals or cells, but are becoming understood and accepted by many regulatory administrations. Furthermore, there is widespread use of materials which have been shown to be accepted by the human body. Also, the EN 550-2-4 standards on sterilization (CEN 94) permit the reliable assessment of device sterility by control of the sterilization process, rather than the batch testing of the products themselves.

These approaches are not generally available to pharmaceuticals. Although some aspects of pharmaceuticals, such as the preliminary tests for biological safety, can be examined by methods similar to those used for medical devices, the scope of laboratory tests is limited. The 'performance' of pharmaceuticals is only manifest by their effects on human patients and there are no laboratory analogues. The assessment of the safety of pharmaceuticals is based on a progression through animal tests to controlled clinical trials, involving a variety of dosage regimes and a search for side effects and contra-indications. Of course, the same techniques have to be used in cases of new devices producing an unknown effect. This fact does not invalidate the argument that laboratory testing against standards is an efficient and effective method of assessing the safety of many medical devices. Another feature of laboratory physical testing is that it generally permits 'overstressing' of the device, i.e. examination of the safety and performance of the device under unrealistically severe conditions such as at high temperature and/or humidity, with power supply errors and with artificially induced faults in the device. Satisfactory tests carried out under such unreal conditions are viewed as giving extra confidence in the device.

The advantages of basing legislation on reference to standards can also be clearly seen. The first and most obvious is the flexibility attached to the technical details. Older legislation which included technical details in the legislative texts were prone to obsolescence as it is generally difficult to make changes in legislation. 'Old Approach' EC Directives frequently referred to a committee set up to adapt the Directive to technical progress as a means of avoiding amending the Directive itself but this, although a useful expedient, still leaves manufacturers with a rigid set of requirements as they are an integral part of the legal instrument.

Restricting the legal obligations to compliance with essential requirements allows the standards (which are not themselves mandatory) to be updated in the normal fashion (the ISO/IEC Directives (Part 1, paragraph 2.9.1) require every international standard to be reviewed at least every five years). Suitable framing of the essential requirements must be accomplished for the system to work properly: they must be sufficiently detailed to allow compliance with them to be determined directly (it is not sufficient merely to require the product in question to be safe) in cases where the harmonized standards are not used, but must not be so detailed that they usurp the role of the standards. The standards then written must address directly and clearly each of the essential requirements and must include test methods and pass/fail criteria.

One of the main objections to the principle of reference to standards is that the regulators thereby hand over their prime responsibility for the safety of patients in their countries to voluntary bodies which themselves do not carry any responsibility. This objection is overcome if compliance with well-framed essential requirements is made the legal obligation and if the regulatory authority retains the right to declare whether any standard carries the 'deemed to satisfy' attribute.

A second advantage is that standards, properly written, enable compliance to be determined objectively by any body with appropriate technical facilities and expertise. The regulatory authority can therefore delegate the task of determining compliance to suitable bodies and does not have to carry out the process itself. The 'Monti Report' (EC 97d) states that more than 600 certification bodies, dispersed across the Community, have been notified to date by the Member States. In fact, the competence and diligence of the Notified Bodies in carrying out their duties in respect of medical devices has been the subject of criticism (Huriet 1996, Kent 1996) and, consequently, more stringent criteria for the selection of Notified Bodies may be imposed. These teething troubles being experienced in the EC do not, however, invalidate the argument that delegation is possible—even to the extent of the acceptance of manufacturers' own declarations of conformity.

A third advantage is that the regulatory authorities can press for new standards, or revisions of existing standards, to address problems reported under a 'vigilance' scheme (see Chapter 9). Despite the time taken to produce standards, this process is generally much quicker than that of amending detailed technical legislation. (In the author's experience, the UK Department of Health made a common practice of feeding information about defective products into the BSI standards committees and, later, into international standards bodies.)

The recognition by leading regulatory authorities participating in the Global Harmonization Task Force (GHTF) (see Chapter 10) of the value of standards in assessing the safety of medical devices has been shown by the development of two GHTF documents:

- Essential Principles of Safety and Performance of Medical Devices (included as Appendix 2)
- Role of Standards in the Assessment of Medical Devices (included as Appendix 3).

These two documents relate very closely to the way in which standards are used in the EC medical device Directives. If ratified by the GHTF, the principle of reference to standards is likely to be given more prominence by GHTF members and to become more widespread in countries influenced by the GHTF.

Commentary

The principle of reference to standards is well established in the European Community; the acceptance or reproduction of the EC medical device Directives in countries such as Australia, Canada and several East European countries has given impetus to the principle, and the opposition often expressed by the US Food and Drug Administration has been partly overcome by the passage of the FDA Modernization Act of 1997 (USA 97a).

Although the Medical Device Amendments of 1976 specified that compliance with standards was the criterion for approval of Class II devices, the reservation was that the standards in question were to be produced by the FDA. The process of standards development proved to be too difficult, no standards were adopted, and the 510(k) 'substantial equivalence' procedure became routine but standards came steadily into use in internal guidance on the pre-market approval procedure.

The FDAMA, which is discussed more fully in Chapter 4, includes the listing of standards and the acceptance of manufacturers' declarations of conformity as satisfying the regulatory requirements in certain circumstances. The FDA is party to the publications by the Global Harmonization Task Force on 'Essential Principles of Safety and Performance for Medical Devices' and the 'Role of Standards in the Assessment of Medical Devices'. The promulgation of these documents is likely to encourage further use of reference to standards by the US authorities and by many other countries.

The former reluctance of the United States, which is an acknowledged leader in medical device legislation, to adopt the principle of reference to standards has no doubt been a factor influencing other countries which take their lead from the US and this recent change of position, together with the backing of the Global Harmonization Task Force, suggests that there is now a clear global trend towards medical device legislation based on this principle.

Chapter 8

The question of effectiveness

Introduction

It is widely accepted that 'safety', i.e. 'freedom from unacceptable risk',[1] alone is not a sufficient criterion for the legal marketing of medical devices; the devices must be capable of operating in some way that contributes to the diagnosis, therapy or support of the patient. This additional feature is certainly a key factor in the regulation of medical devices. It is described in different ways, but is often referred to as 'effectiveness', 'efficacy' or 'performance'. This Chapter examines the differences between these concepts and how devices are determined to satisfy the requirements, and proposes a model for future legislation.

Legislative requirements

US requirements

The US regulations (see Chapter 4) are intended to 'provide a reasonable assurance of the safety and effectiveness of the device' (USA 76a, Sec. 513), where the safety and effectiveness are to be determined:

'(A) with respect to the persons for whose use the device is represented or intended,
(B) with respect to the conditions of use prescribed, recommended, or suggested in the labeling of the device, and
(C) weighing any probable benefit to health from the use of the device against any probable risk of injury or illness from such use.'

---

[1] As defined in ISO/IEC Guide 51: Safety aspects—Guidelines for their inclusion in standards (ISO/IEC 99b).

and where the effectiveness of a device is 'to be determined … on the basis of well-controlled investigations, including 1 or more clinical investigations where appropriate, by experts qualified by training and experience to evaluate the effectiveness of the device, from which investigations it can fairly and responsibly be concluded by qualified experts that the device will have the effect it purports or is represented to have under the conditions of use prescribed, recommended or suggested in the labeling of the device.' (USA 76a, Sec 513 (a) (3) (A).)

Japanese requirements

The Japanese regulations (see Chapter 4) are intended to 'assure the quality, efficacy and safety' of medical devices (Japan 94a, Article 1). No definitions of 'quality' or 'efficacy' are given. Article 14, which requires approval to be given for the manufacture of drugs, devices etc., goes on to say in paragraph 2 'The approvals specified in the previous paragraph shall be based on the examination of the name, ingredients, quantities, structure, directions, dosage, method of use, indications, effects, performance, adverse reaction, etc. of the drug, quasi-drug, cosmetic or medical device concerned. Approval shall not be granted when any of the following conditions are met:

'(1) The drug, quasi-drug or medical device is not shown to possess the indications, effects or properties indicated in the application.
(2)   The drug, quasi-drug or medical device in the application is found to have no value as a drug, quasi-drug or medical device because it has harmful action which outweighs its indications, effects and properties.'

EC requirements

In the EC Medical Devices Directive (MDD) (EC 93a) the words 'effectiveness' and 'efficacy' are not used. The requirement is expressed in the essential requirements (Annex 1 of the Directive), three of which must be considered together:

'1. The devices must be designed and manufactured in such a way that, when used under the conditions and for the purposes intended, they will not compromise the clinical condition or the safety of patients, or the safety and health of users or, where applicable, other persons, provided that any risks which may be associated with their use constitute acceptable risks when weighed against the benefits to the patient and are compatible with a high level of protection of health and safety.
3.   The devices must achieve the performances intended by the manufacturer and be designed, manufactured and packaged in such a way that they are suitable for one or more of the functions referred to in Article

1 (2) (a) as specified by the manufacturer.' [Article 1 (2) (a) is the definition of a medical device (see Chapter 1)].

6. Any undesirable side-effect must constitute an acceptable risk when weighed against the performances intended.'

Understanding the requirements

To understand these requirements, and the difference between 'effectiveness/efficacy' and 'performance', it is necessary to review not only the wording of the legislation but also other evidence relating to its interpretation. Such evidence may be found in definitions, in records giving the background to the legislation, and in requirements and guidance for the conduct of clinical trials carried out to provide the evidence for the effectiveness or the performance of medical devices. In this section the definitions and background will be reviewed.

Definitions

The requirements in the US regulations refer to 'effectiveness'; the Japanese regulations refer to 'efficacy'. These two terms have been defined as follows (Last 1988):

Effectiveness: The extent to which a specific intervention, procedure, regimen or service, when deployed in the field, does what it is intended to do for a defined population.

Efficacy: The extent to which a specific intervention, procedure, regimen or service produces a beneficial result under ideal conditions.

'Effectiveness' appears to be the more appropriate consideration for medical device legislation, which is aimed at protecting the public under all conditions, and, in the absence of a definition in the Japanese regulations, and in view of the uncertainty of the translation from the Japanese original, it will be assumed in the following discussion that both sets of regulations refer to effectiveness.

Although there is no definition of performance in the Directives, the definition given in EN 540: Clinical investigation of medical devices for human subjects (CEN 93) can be used:

Performance: The action of a specific medical device with reference to its intended use when correctly applied to appropriate subjects.

This definition does not refer to the results of the action of the device (i.e. the outcome) and it is this which encapsulates the difference between effectiveness and performance. It can be argued that the outcome, resulting from the use of a device, may be influenced by other factors such as the accuracy of diagnosis, other treatments being applied, the skill of the doctor, and even the particular characteristics of the patient(s). Performance is regarded as being a true characteristic of the device alone which can be determined objectively.

Background—USA

Most of the available background comes from the United States where 'effectiveness' was introduced in the Medical Device Amendments of 1976 (USA 76a) and has been the subject of some controversy in the succeeding years. In the drafting of the Amendments, Congress recognized 'that no regulatory mechanism can guarantee that a product will never cause injury, or will always produce effective results. Rather, the objective of the legislation is to establish a mechanism in which the public is afforded reasonable assurance that medical devices are safe and effective' (USA 76b). The Congress Report goes on to say 'Devices vary widely in type and in mode of operation, as well as in the scope of testing and experience they have received. Thus, the Committee has authorized the Secretary to accept meaningful data developed under procedures less rigorous than well-controlled investigations in instances in which well-documented case histories assure protection of the public health or in instances in which well-controlled investigations would present undue risks to subjects or patients.'

Hutt *et al.* (1992) conclude that this Report directly acknowledged the concerns expressed in the Congressional hearings that a drug-type standard of evidence for effectiveness was inappropriate for devices. In discussing the evidence presented at the hearings, they quote Mr Kenneth Marshall as explaining 'drugs working by chemical action tend to work on their own, independent of an operator, and so efficacy there [i.e. in the new drug context] really reads more that the device [*sic*] must not only work but perform ... A device, on the other hand, is really an extension of the user's skills ... A device can be no more effective than the user of the device, who is seldom the patient and most frequently the health professional.' (USA 73), and go on to say that, accordingly, Mr Marshall suggested that evidence of 'efficacy' in the device context amounts to proof that the device will perform as intended, not that it will necessarily produce therapeutic effects.

It appears that these views were never accepted in practice. The internal guidance used in the Center for Devices and Radiological Health for the review of pre-market approval applications (PMAs) (USA 86) states 'The PMA applicant must provide a cogent demonstration of the safety and effectiveness for all diagnostic and/or therapeutic medical claims for the device based on the data and analyses in the laboratory, animal, and clinical data sections. It is necessary that the study protocol, results, analyses and interpretation support and be consistent with the medical claims for the device.'

In describing the clinical data requirements, it states that a statistically valid clinical investigation should include 'a comparison of the results of treatment with a control (including historical control) or standard to permit quantitative evaluation. Generally, four types of comparison groups are recognized: (1) no treatments, (2) placebo control, (3) active treatment control, and (4) historical control. Historical controls are the weakest type of controls from a statistical viewpoint because it is very difficult to assure comparability of important prognostic variables with the treated group, especially if the disease or therapy has changed over time'.

This appears to be much closer to the process of establishing the efficacy of a drug[2] (see, for instance, Spilker (1991) pp. 166–168) than to the approach advocated by Marshall and others. The 'drug approach' was further strengthened by the issue of the 'Temple Report' (USA 93c) in March 1993. This was the report of the Committee for Clinical Review, set up 'to provide (the Center for Devices and Radiological Health) with recommendations on improving design and performance of clinical studies to be submitted in support of device applications.' The Committee was composed of staff members from the Center for Drug Evaluation and Research (CDER) and was chaired by Dr Robert Temple, Director of the Office of Drug Evaluation.

The Committee found that there were significant deficiencies in the design, reporting and analysis of most clinical studies included in PMA applications. Consequently, 'the result is likely to be a device whose performance is less well-characterized than would be desirable, with less information on the utility and performance of the device than could have been obtained with only modestly greater, but better directed, effort. For example, it was impossible in some cases to determine whether a device intended for a life-threatening condition was as effective or safe as already available therapies, *an important lack of information for the physician and patient who must decide whether to choose the new device over available alternatives*' (author's emphasis).

The Committee thus took the view that it is the task of the regulatory authorities to compare a new device with alternative methods of diagnosis or therapy and to approve it only if is it shown to be as good as, or better than, available alternatives. The author has not found any other regulatory authority which accepts this task as its responsibility—it is generally seen as an issue for the medical profession, reimbursement agencies or other government institutions.[3]

Holstein and Wilson (1997) comment (p. 282) 'The House Report (USA 76b) explains that the FDA can make use of data developed under procedures that are not as rigorous as those for drugs. The language in

---

[2] 'Efficacy is called for in most drug legislation, e.g. EC Directive 65/65/EEC (EC 65).

[3] The UK government has announced (UK 97b) the formation of a National Institute for Clinical Excellence (NICE) charged with promoting clinical- and cost-effectiveness.

section 515 of the Food Drug and Cosmetic Act clearly allows the FDA to rely on less formal evidence to demonstrate safety and effectiveness than the agency has prescribed for new pharmaceuticals. Nonetheless, in many cases, the FDA requires studies on medical devices that are designed like new drug trials (e.g., prospective, randomized controls). This requirement is based in part on the results of the report of the Committee for Clinical Review (also known as the Temple Report (USA 93c)) in 1993. The Committee accepted the proposition that "the fundamental principles underlying evaluation of any therapeutic intervention, whether it is a drug, device, diet or surgical procedure, are the same".'

This resulted in a very drug-orientated approach to the effectiveness of medical devices. Kahan (1995) quotes two cases where the FDA took what he regards as extreme positions, going well beyond his view of what is required for the release of products on to the market:

'In a recent example, the sponsors of a clinical study for a device intended for treating hypercholesterolemia were able to clearly demonstrate that the device reduced total cholesterol or total low density lipoproteins. The device was effective for its intended use. However, to demonstrate clinical utility, FDA required that regression studies show a reduction in the formation of arthrosclerotic plaque in coronary vessels, which FDA believed would indicate a reduced risk of heart attack.

Another example involves . . . an assay for a serum tumor marker for pancreatic cancer. The sensitivity of the assay was shown to be the greatest in the late stages of disease. The (advisory) panel felt that this product was not a useful improvement over other alternative methods of diagnosis, since a positive result frequently occurred too late to intervene in the course of the disease; therefore, the panel found that the assay had no clinical benefit.'

This approach to establishing effectiveness attracted some unfavourable comments. Munsey (1995, p. 177) writes 'Even though the MDA (Medical Device Amendments) contained many provisions reflecting the views of Congress that devices and drugs were to be treated differently, CDRH (the Center for Devices and Radiological Health) is beginning to regulate devices in ways similar to the ways CDER (the Center for Drug Evaluation and Research) regulates drugs.'

The US medical device manufacturing industry reacted against this trend and the 'Wilkerson Report' (Wilkerson 1995, p. 20) attributes several problems to the length and extent of the approval process in the US, particularly the review of clinical utility: 'The agency is expanding its review beyond its traditional measures of safety and performance to now include efficacy and clinical utility. These changes have led to long delays in approving products by the agency.' The Report concludes that, as one of the consequences of these delays, 'patients in the US are gaining access to important advances in medical technology months, even years, later than patients in other countries' and 'Delayed US patient access to new technology reflects lost opportunities to save lives, reduce hospitalization and recovery time, lower complication and infection rates, and improve patient quality of life.'

Pressure of this kind resulted in consideration by Congress of amending legislation which would address several concerns about the activities of the Food and Drug Administration. Pilot and Waldman (1998, p. 273) observe 'Of particular concern were perceived attempts by FDA to evaluate the cost-effectiveness of new devices, to require devices to prove themselves more effective than alternative therapies already on the market, and to require evidence of clinical utility.'

The resulting FDA Modernization Act of 1997 (USA 97a) included, in section 205, provisions requiring FDA, in evaluating a PMA application, to rely solely on the conditions of use submitted in the labelling; to consider the extent to which reliance on post-market controls might reduce the amount and type of data needed to support the application; and to consider the least burdensome means of evaluating effectiveness that would have a reasonable likelihood of resulting in approval. There is not yet evidence of the effects of these changes in the regulations.

Background—Japan

There is no information available about the background to the Japanese approach to the establishment of medical device effectiveness. The requirements in Japan are explained in guidance published by the Ministry of Health and Welfare (Japan 94c, pp. 139–141). In discussing the clinical trials results that should be included in an application for manufacturing approval of a medical device, it states: 'Results of clinical trials are very important in that they provide information on expected effects and adverse reactions in the clinical use of the medical device under application, and that they afford a right key to the evaluation of the effectiveness of the device ... First of all, an opinion of a physician in charge of a clinical trial must be presented in the report, which is the most important factor in the evaluation of the device ... To the report of clinical trial results, a list of case records shall be attached, describing ... improvement of general conditions, adverse reactions and the outcome, exclusions and dropouts and the reasons, and availability (overall evaluation) of the device.' This guidance is certainly aligned with the US requirements but there is much less detail given on the set-up and statistical validity of the trial. This must reflect an expectation of much less far-reaching conclusions to be drawn from the trial than would be expected in the US. Although mention is made of comparison with a control, or dummy, device in the examples given in the MHW guidance (pp. 157–164), there is no suggestion that a device must be shown to have equivalent diagnostic or therapeutic possibilities as competing devices as advocated in the Temple Report.

Background—EC

As discussed in Chapter 2, the first drafts of the Directives on active implantable devices and on medical devices were prepared by industry groups. Alan J Howard, a lawyer who was a principal drafter of both texts, has stated (Howard 1998) that European industry was totally opposed to the American approach to effectiveness and, from the beginning, insisted on the word 'performance' appearing in the texts.

The industry view was challenged during meetings of the Commission's Working Group on Medical Equipment, and Doolan (1989) records the following discussion at the meeting of 15/16 June 1989:

> 'There was much discussion of a need to evaluate the efficacy of medical devices.
>
> At this meeting, it was clear that much confusion exists in relation to these terms and clear definitions of performance and efficacy are needed for the next round of discussions. The UK (H Sutton) felt efficacy was important, but defined it like performance.
>
> Germany (G Schorn) spoke of the need for evaluation to show therapeutic or diagnostic effectiveness and to make observations of side-effects being needed for devices through clinical trials. UK (M N Duncan) indicated that clinical evaluation should be required and suggested that "efficacy" was "proven effectiveness in skilled hands". Germany felt that, in some cases, clinical evaluation could be replaced by reference to earlier tests or to the literature. France (E Waisbord), UK and Germany expressed the view that some form of clinical evaluation would be needed for active non-implantable medical devices and for some non-active medical devices.'

These rather vague and somewhat conflicting views were reconciled, by writing Section 3 of the essential requirements[4] to address performance, and adding Sections 1 and 6 which require consideration of risks and side-effects and a risk/benefit analysis if risks and/or side-effects are associated with the device.

These two requirements which, according to Annex X, must 'as a general rule' be verified, in addition to the performance itself, by the evaluation of clinical data bring the European requirements appreciably closer to those of the US and Japan than may be generally realized.

---

[4] Doolan also records that at this time the 'essential safety requirements' formerly included in the draft Directive on Active Implantable Medical Devices were changed to 'essential requirements', reflecting the fact that aspects other than safety had to be included.

The clinical evaluation of medical devices

Considerable light is thrown on the issue of 'performance' versus 'effectiveness' by examination of the requirements for clinical evaluation and clinical investigation in the three administrations.

The difference between these two terms must be understood: a **clinical evaluation** means an evaluation of available evidence (data) on the safety and performance of a medical device in its intended clinical use; a **clinical investigation** (clinical trial) is a means of obtaining such evidence (data).

Clinical evaluation in the USA

The US requirements quoted in Section 9.2.1 apply generally to all devices other than those in Class I. Exceptions are (USA 76a, Sec.513(a)(3)(B)):

'If the Secretary determines that there exists valid scientific evidence (other than evidence derived from investigations described in subparagraph (A)) (i.e. clinical investigations)
    (i) which is sufficient to determine the effectiveness of the device, and
    (ii) from which it can fairly and responsibly be concluded by qualified experts that the device will have the effect it purports or is represented to have under the conditions of use prescribed, recommended, or suggested in the labeling of the device,
then, for purposes of this section and sections 514 and 515, the Secretary may authorize the effectiveness of the device to be determined on the basis of such evidence.'

The net effect of these requirements was that for devices approvable under the 510(k) procedure the establishment of 'substantial equivalence' was regarded as providing valid scientific evidence of the effectiveness of the device and clinical trials were not generally required. With the passage of time, and particularly following the publication of the Temple Report, clinical data began to be requested frequently in 510(k) submissions. The FDA Modernization Act of 1997 has applied a correction by the addition of section 513(a)(3)(C)(ii) 'Any clinical data, including one or more well-controlled investigations, specified in writing by the Secretary for demonstrating a reasonable assurance of device effectiveness shall be specified as a result of a determination by the Secretary that such data are necessary to establish device effectiveness. The Secretary shall consider, in consultation with the applicant, the least burdensome appropriate means of evaluating device effectiveness that would have a reasonable likelihood of resulting in approval.'

These changes should have the effect of restricting the demand for clinical trials to cases where pre-market approval is necessary and where the novelty of the device is such that there is no available evidence of its effectiveness.

Clinical evaluation in Japan

In Japan, guidance from the Ministry of Health and Welfare states (Japan 94c, p. 140):

> '...clinical study data shall be submitted for the products under the following three conditions:
> a. New medical devices in Japan.
> b. Already approved medical devices whose purpose or scope of use is to be extended.
> c. Already approved medical devices whose safety is the most important factor.'

Unlike the situation in the USA, described above (and in the EC—see below), where data from the literature, relating to essentially similar devices, may be considered, the Japanese regulations state specifically that anyone seeking approval shall attach 'data concerning the results of clinical trials and other pertinent data to their application' (Japan 94a, Article 14.4), inferring that trials must always be carried out on the devices in question although, as the categories listed above are quite restricted and likely anyway to go beyond information in the literature, this does not seem to be a more burdensome demand than would be met elsewhere.

Clinical evaluation in the EC

Annex X of the MDD addresses clinical evaluation and states:

> '1.1   As a general rule, confirmation of conformity with the requirements concerning the characteristics and performances referred to in Sections 1 and 3 of Annex I under the normal conditions of use of the device and the evaluation of the undesirable side-effects must be based on clinical data in particular in the case of implantable devices and devices in Class III. Taking account of any relevant harmonized standards, where appropriate, the adequacy of the clinical data must be based on:
> 1.1.1 either a compilation of the relevant scientific literature currently available on the intended purpose of the device and the techniques employed as well as, if appropriate, a written report containing a critical evaluation of this compilation;
> 1.1.2 or the results of all the clinical investigations made, including those carried out in conformity with Section 2.'

Section 2 describes the conditions under which clinical investigations must be carried out. The corresponding harmonized standard is EN 540: *Clinical investigation of medical devices for human subjects* (CEN 93). This standard offers its own definition:

Clinical investigation: any systematic study in human subjects, undertaken to verify the safety and performance of a specific medical device, under normal conditions of use.

'Safety' in the definition above clearly includes issues of any risks or side-effects presented by the device, and these are the only aspects of the investigation which directly depend on a study of patient outcome. 'Performance' is seen as a characteristic of the device alone, to be determined when it is correctly used for its intended purpose on patients, but which is not assessed by a study of patient outcome.

Where risks or side-effects are presented by the device, risk/benefit analysis is required, but there is no question of comparing devices or of assembling information for the physician or patient to enable them to choose one device rather than another (as proposed by the Temple Report).

Although Annex X of the MDD refers to clinical evaluation being carried out 'As a general rule', most medical devices do not present risks or side-effects which call for risk/benefit analysis, and their performance can be determined in the laboratory. The situation is, therefore, that clinical evaluation is, in practice, limited to the implantable and Class III devices identified in Annex X plus other categories of device suggested by the European Notified Bodies Group (NB 98a) (see Chapter 3):

- the introduction of a device using a novel technology whose features and/or mode of action are unknown
- where a device incorporates new materials coming into contact with the body or known materials being used in a new way or in a new location
- where a potentially hazardous technology is used for a new indication
- when a device approved via clinical evaluation is significantly modified.

These categories appear to be more restricted than those for which clinical data are required in the United States, but the amendments introduced by the FDAMA are likely to restrict US requirements in much the same way.

The conduct of clinical investigations

Section 520(g) of the Federal Food, Drug and Cosmetic Act allows medical devices to be exempted from the normal approval requirements 'to permit the investigational use of such devices by experts qualified by scientific training and experience to investigate the safety and effectiveness of such devices.' The section goes on to describe in outline the conditions under which the clinical investigation must be carried out, including the submission of an investigational plan to an institutional review board (corresponds to an ethics committee) or to the FDA, informed consent of the patients, and appropriate record keeping. Detailed guidance on clinical trials has been published elsewhere (USA 86).[5]

The details of the proposed clinical trial form an important part of the submission for Investigational Device Exemption and, in the years since the 1976 Medical Device Amendments were introduced, the FDA has formed views on the size of clinical trial for many types of device and has made them known in guidance to its staff and to manufacturers.[6]

Article 80-2 of the Japanese Pharmaceutical Affairs Law (Japan 94a) requires persons commissioning clinical trials to obtain data for inclusion in an application for manufacturing approval to ensure that the trials are carried out on the basis of standards laid down in MHW ordinance. Paragraph 2 of the Article requires persons commissioning clinical trials to submit the plan of the trials to the Ministry which may cancel or modify the trial.

The conditions for approval of a clinical trial (which are the same for drugs and devices) are specified in Articles 67, 68, 69 and 70 of the Enforcement Regulations. The conditions regarding submission of results of pre-clinical tests, details of the investigators and institutions, the trial protocol, etc. are generally similar to those in Europe and the US. Trials on medical devices are to be conducted in accordance with Good Clinical Practice for Trials on Medical Devices (Japan 92a) and GCP Manual for Medical Devices (Japan 93). Again, these documents enforce conditions similar to those in the other countries considered, including formal contracts between the sponsor and the institution; approval by an institutional review board; informed consent by patients, and reporting of adverse events. More details are given in the Guide to Medical Device Registration in Japan (Japan 94c), Chapter 9, pp. 227–340. Japan is similar in many ways to the USA in the degree of control exercised; the MHW has specified the size of trial required for 21 types of device in its published guidance (Japan 94c).

Article 4 of the EC Medical Devices Directive allows devices which do not bear the CE marking to be supplied for clinical investigation 'if they meet the conditions laid down in Article 15 and Annex VIII'. These conditions involve obtaining the agreement of an ethics committee and notification to the Competent Authorities of the countries where the trial is to be carried out.

The trial must be conducted in accordance with Annex X of the Directive. This requires the preparation of an investigational plan, use of appropriate procedures, testing under normal conditions of use, examination of all appropriate features, recording of adverse incidents, conduct under the responsibility of a suitably qualified person, and the preparation of a

---

[5] For a fuller description of the US requirements see Kahan (1995) or Segal (1998).
[6] See, for instance, the FDA Medical Device Clinical Study Guidance (USA 93d), Draft Replacement Heart Valve Guidance (USA 93e).

signed report of the investigation. This Annex also requires all clinical investigations to be carried out in accordance with the Helsinki Declaration (Helsinki 64).

The Member States have implemented these basic requirements in their national regulations with some flexibility and, in practice, manufacturers wishing to carry out clinical investigations—which are often needed before the CE marking can be applied—have to take care to comply with the national procedures.

The requirements for the conduct of the clinical investigations have been described in detail in EN 540 '*Clinical investigation of medical devices for human subjects*' (CEN 93) which is the harmonized standard corresponding to Annex X. Although this is a voluntary standard (as are all standards in the European system), it is hard to imagine any clinical trials being mounted on any other basis.

In Europe, the assessment of the data is the province of the Notified Bodies. With some 60 NBs, there is bound to be some uncertainty in this process. Some NBs have encouraged medically qualified personnel to join their staffs but, in most cases, the Notified Body engages a clinical expert to assist in the assessment on a product-by-product basis. Recent experience of several Notified Bodies (Anderson 1997; Horn 1998) suggests that NBs are ready to accept data from the literature whenever possible, are making somewhat superficial assessments of performance, and do not yet have the experience or confidence to be more demanding of the submitted clinical data or to make outcome studies for risk/benefit analysis.

The Notified Bodies Group has recently issued guidance on the evaluation of clinical data (NB 99). Adherence to this guidance will regularize the approach, and improve the consistency, of Notified Bodies in carrying out clinical evaluations but further development of this document is likely to be necessary before it can be considered as resolving all the difficulties in this process.

In the USA and Japan, the assessment of clinical data and decisions on approval of devices are reserved to the administrations.

There is a very close similarity, in the three administrations considered in this chapter, in the approval processes leading to the initiation of a clinical investigation and in the conditions under which investigations are to be carried out. This has resulted in a growing acceptance of clinical data from trials carried out in a country other than the one where submission is made.

Discussion

The assessment of medical devices under any regulatory system involves two elements: safety and effectiveness/performance. These two elements may be considered separately.

As argued in Chapter 7, the hazards presented by many medical devices are known physical hazards which can be checked by laboratory testing. However, other devices, particularly implants, devices which impart energy to the patient, and novel devices, present possible unknown hazards and side-effects which can only be determined by careful examination of their use on patients.

Despite the detailed differences in the regulations examined earlier in this chapter, there is a large measure of agreement that it is devices of these types for which clinical evidence of safety is required. Constructive study by the regulatory authorities should result in agreement on the types of device for which clinical evidence of safety is necessary for approval.

There is no worldwide agreement on the extent of the evidence required, but in all regulatory systems there is an acceptance of historical evidence relating to closely similar devices. There is also a growing understanding that the practical length of a pre-market clinical trial is not sufficient to uncover all possible adverse reactions and side-effects and that long-term follow-up, or post-market surveillance, is necessary if such undesirable effects are to be detected and (if possible) corrected with minimum distress to the population of patients using the device (see Chapter 9).

There is also a requirement (explicit in the US and European regulations; implicit in the Japanese) that, where risks and side-effects are discovered, a risk/benefit assessment must be carried out. There is no general understanding of how such assessments are to be made and few examples of risk/benefit assessments. It may be noted that X-ray diagnosis, which carries some risk, is generally accepted as conferring benefits which outweigh the risk although it should be noted that the UK guidance on the medical use of X-rays (UK 87b) states (Sections 2.3 and 2.4) 'All diagnostic or therapeutic procedures including exposure to radiation for medical purposes may carry some personal risk.... It is important therefore that only those medical exposures that are necessary should be undertaken.... A person who requests an examination should be satisfied that it is necessary, taking into consideration the benefits expected from the examination and the radiation dose involved' and, in Section 2.16, 'Screening programmes, e.g. chest radiographs, should be undertaken only if the expected medical benefits to the individuals examined and to the population as a whole exceed the economic and social costs, including the risks associated with the radiation dose involved. Since benefits are not always the same for all members of the population, screening should be limited normally to particular groups.' It is noteworthy that one of the few published examples of a risk/benefit analysis is contained in the Forrest Report on breast screening by mammography (UK 86) which concluded that deaths from breast cancer for women between the ages of 50 and 64 could be reduced by one third or more, and that 'the radiation risk of regular mammography for screening women over 50 years old is thus insignificant compared with its potential benefits' (UK 86, Chapter 2, paragraph 2.18).

Although the place of clinical data and clinical investigations in the assessment of safety is reasonably consistent in the three administrations considered, the same cannot be said for the assessment of effectiveness/performance.

It is accepted that most 'new' medical devices are incremental improvements of existing products. In such cases the effectiveness of the device, which the author interprets as the effect the device has on the outcome for the patient, is already known from experience but the performance, which the author interprets as a quantitative description of the action of the device, may have been changed and should be confirmed. For most devices this may be done in the laboratory under conditions simulating its use on the patient but for some only actual use on the patient gives the true conditions.

For implanted devices it is generally impossible to make quantitative measurements during their use on patients, but clinical experience is needed to confirm laboratory measurements. For instance, a new artificial heart valve may show improved fluid flow characteristics in the laboratory but quantitative measurements are difficult in vivo and the performance may be confirmed by measurements of blood haemolysis (which the improved flow is intended to reduce) and of the patient's general condition, and by the patient's own feeling of well-being; i.e. patient outcome may be used as a confirmation of performance. Grunkemeier (1998) states 'However, the effectiveness of a medical device (its adequacy to perform its mechanical function) can be tested using in vitro studies and in animals, and can be measured directly in humans. In the case of a heart valve, this function can be measured by means of such parameters as gradient, cardiac output, transvalvular regurgitation, and so forth, and can even be inferred to a large extent by a lack of symptoms.' It is interesting to see that Grunkemeier interprets effectiveness as being synonymous with performance and regards the outcome, i.e. elimination of symptoms, merely as a confirmation of the performance of the device.

Another example is that of a new artificial hip joint showing reduced friction when measured in the laboratory but which depends on a subjective assessment by the patient for confirmation that this is experienced after implantation. These examples demonstrate the interplay between performance and effectiveness, and the difficulty of separating them in the case of implants, but there is a basic difference between observing patient outcome as a confirmation of measured performance, or in order to allow a risk/benefit analysis, and the conduct of a clinical investigation to determine the effectiveness of a device by comparison with an alternative device or procedure or by comparison with placebo.

Van Vleet (1998) points out that 'Orthopedic device studies, by nature of their delivery systems, preclude the possibility of using placebo controls.' He makes some interesting comments on the assessment of patient outcome as a determinant of device effectiveness: 'Many ambitious clinical scientists attempt to incorporate outcome evaluations into clinical trials. The controversy surrounding the clinical utility of patient satisfaction may encourage companies to amass patients' psychosocial data to substantiate their claims of successful treatment. These evaluations may include everything from Visual Analogue Scale (VAS) determinations of patients' "quality of life" to their ability to walk their pets or play golf. At the risk of undermining the importance of this new and interesting area of clinical research, it is critical to emphasize that measurement of outcomes have applications in clinical trials *only* when used with already validated parametric tools . . . '.

Both Grunkemeier and Van Vleet comment on the inability of clinical trials of any practical size and duration to give a complete picture of the safety and performance/effectiveness of long-term implants. Events such as the fatigue failure of the Bjork-Shiley heart valve or loosening of hip implants occur after long periods, cannot be detected by pre-market assessments, and need other approaches to their detection and alleviation—a subject that is discussed further in Chapter 9.

Commentary

The clinical evaluation of medical devices is probably the least well understood aspect of the approval process. Despite repeated assertions that devices are not drugs (USA 73, Hutt *et al.* 1992, Kahan 1995, Holstein and Wilson 1997, Witkin 1998a), both the US and Japanese authorities treat devices in ways very similar to the way in which they treat drugs.

Only in Europe has an approach to medical devices, clearly distinct from that of drugs, developed. This concentrates on establishing safety and performance wherever possible on the basis of laboratory tests and historical data relating to essentially similar devices, and embarking on clinical investigations only in cases where these are necessary. The European process is still new and needs considerable development, particularly in establishing the size and scope of clinical investigations and in producing guidance on the assessment of clinical data, but it seems to be a more relevant approach to the protection of the public health—when coupled with post-market surveillance—than the US concentration on comparisons of outcomes from alternative devices. As Witkin (1998b, pp. 79/80) states in her introduction to a number of clinical case studies, 'These chapters illustrate that a range of clinical study designs can be relied on to generate valid data for use in formulating a sound risk/benefit analysis of device performance within the context of existing clinical practice . . . . Hence, as these authors note, there is and should be a continued reliance on extensive nonclinical research (bench testing and animal models) in the evaluation of device performance.'

The European approach is another instance of the progress, which is being made and needs to be continued, from treating devices as a special kind of drug to one of treating them as engineering products. The FDA Guidance for Clinical Investigations for a PMA (USA 86) states (p. III-9) 'The PMA applicant must provide a cogent demonstration of the safety and effectiveness for all diagnostic and/or therapeutic medical claims for the device...it is necessary that the study protocol, results, analyses and interpretation support and be consistent with the medical claims for the device.' This is a very drug-like phraseology, as is Article 14.2 of the Japanese Pharmaceutical Affairs Law which makes no distinction between drugs and devices.

Nevertheless, from this study it is clear that, where clinical data are required for decisions about safety, performance and risk versus benefit, the categories of device in question are much the same in the three administrations studied and that discussions between them could probably result in a common description of the device categories. Only the US, with its demand for outcome studies to demonstrate comparative effectiveness, has appreciably wider requirements for clinical studies and this may be reduced as the FDAMA comes into effect.

Furthermore, there is some consistency in the conditions under which clinical trials may be carried out. In all three administrations, permission is needed before an unapproved medical device may be used on patients to gather the information needed for its approval. In all cases it is necessary to show that all pre-clinical tests have shown the device to be safe; the trial must be carried out under conditions generally reflecting the Helsinki Declaration (Helsinki 64), and the arrangements for the trial must be agreed by the authorities—usually on the basis of approval by an ethics committee or an Institutional Review Board.

These conditions are essentially identical; the variations within the EC may be as great as those between Europe and the other administrations. Differences lie mainly in the number of patients/centres involved and the degree of scrutiny given to the protocol.

There remains the question of the objective of the trial. As discussed at length in earlier sections of this chapter, this always includes safety but varies between performance and effectiveness. Where 'effectiveness', without qualification, is required by the legislation, it is difficult for the regulatory body to put limits on the studies. The author has had experience of clinical trials of inordinate length, attempting to satisfy the FDA of the effectiveness of devices for physiotherapy and wound healing. As the effects of such devices are subtle and largely subjective, adequate evidence is difficult to assemble but, as there is no safety issue involved, the public health is not well served by such long-term studies.

Although comparative information about the outcomes obtained from various competing devices with given populations is of undoubted value, this constitutes health technology assessment, defined by Russell and Grimshaw (1995) as 'the evaluation of the benefits and costs (clinical, social, economic and system-wide) of transferring the technology of interest into clinical practice.' This is an exercise more appropriately carried out by reimbursement bodies and strategic planning authorities. Outcomes studies are necessary for health technology assessment (see, for instance, Fryback and Thornbury (1991)), but are only really essential for the public health where risks and side-effects are possible. Furthermore, to be of real value, such studies need to be carried out over a long period of practical use (Rickards and Cunningham (1999))—much longer than is generally possible in a pre-market examination.

In keeping with the engineering approach adopted by the EC, the view could be taken that the task of regulatory authorities should be to ensure that the design of a new medical device is properly validated—as required by ISO 9001 (ISO 94a) and regulations based on that standard—before the device is released to the market. Validation 'concerns the process of examining a product to determine conformity with user needs' (ISO 94d). ISO 13485 (ISO 96a), sub-clause 4.4.8, which applies this concept to medical devices, says 'As part of design validation, the supplier shall perform and maintain records of clinical evaluations.'

The guidance on design control issued by the FDA (on http://www.fda.gov/cdrh/comp/designgd.html) and adopted with minimal changes by the Global Harmonization Task Force, expands on this requirement as follows: 'Many medical devices do not require clinical trials. However, all devices require clinical evaluation and should be tested in the actual or simulated use environment as a part of validation . . . additional methods are often used in conjunction with testing, including analysis and inspection methods, compilation of relevant scientific literature, provision of historical evidence that similar designs and/or materials are clinically safe, and full clinical investigations or clinical trials.'

'Performance' is an engineering concept which, when linked with ISO 9001, ISO 13485 and the guidance quoted above, is more likely to lead to general agreement and consistent assessment than 'effectiveness', and which gives proper and efficient protection of the public health when accompanied by the requirement for risk/benefit analysis where risks or side-effects are present. Of course this implies that risks and side-effects must be searched for when devices unknown, or with unknown features, are under consideration and this is catered for in the guidance from the Notified Bodies Group (NB 96) on when clinical evaluation is needed.

Agreement is needed on the ways in which the performance of medical devices should be expressed. It should, as far as possible, be expressed in quantitative terms which are characteristic of the device itself and which are not likely to be influenced unduly by interference from the characteristics of the patient or the user (where this is not the patient), and medical claims should be avoided. The performance parameters should, of course, relate to its function as a medical device and, at some point, they will begin to resemble medical claims. It is necessary to appreciate the spectrum of 'new' devices from the incrementally improved to the genuinely novel. As novelty increases, and less is known about the device and its effects, the more an approval authority must depend on clinical data to establish that the device is safe, that it has a medical function, and that the benefits offered by the device outweigh any side-effects or other risks it might present. Thus the clinical evaluation becomes more like that of a drug as advocated in the US and Japan, and the differences between the approach in these two countries and that in Europe begin to disappear.

Chai (2000, p. 66) unfortunately quotes from the GAO Report of 1996 (USA 96a) an example purporting to show the difference between the European and the US assessment of new devices: 'An example used by the GAO illustrates this point—the marketing of an excimer laser in the United States requires the demonstration of the laser's abilities to both excise corneal tissue, and to correct visual anomalies. Under the EU system, the ability of the laser to correct visual anomalies needs verification only if the manufacturer claims such a function. Otherwise, all the manufacturer needs to show to obtain approval for marketing the laser is its capacity to remove tissue.' This is, of course, a misreading of the EC Directive which requires a risk/ benefit judgement of a device such as a laser which clearly presents a risk to the patient. Furthermore, according to the criteria suggested by the Notified Bodies Group, a clinical investigation would be required as the basis of the risk/benefit analysis. The difference between the US and EC procedures, there-fore, would be that the EC motivation would be to establish that 'any risks which may be associated with their use constitute acceptable risks when weighed against the benefits to the patient and are compatible with a high level of protection of health and safety' (MDD, Annex I, Essential Require-ment 1) whereas the US motivation would be not only to establish that the benefits justify the risk but possibly to compare laser correction with other methods of correction.

The author's aim is to emphasize the similarities, and not the differences, in approval procedures in the major administrations, which are often obscured by language and style of the regulations and by failure to appreciate what they mean. A good example of obscured similarities is given in the FDA's comments on proposals to improve the medical device programme (USA 95b). In defence of the assessment of effectiveness, it states: 'A "safety only" system would also be undesirable because the safety of a new device cannot be considered separately from its effectiveness. Since few if any medical products are absolutely safe, judging the safety of a new device becomes a benefit-risk decision—the question is not "Is it safe?" but "Is it safe *enough* in view of its anticipated benefits to the patient?" This means that the device's beneficial outcome must be assessed as *part* of the safety determination. Without such a standard, high risk devices like the initial implantable cardioverter defibrillator products which had a surgical mortality of around 3% might have been judged too dangerous to use, whereas, effectiveness data showed that, on balance, they saved lives.' It is difficult to imagine any regulator, in Europe or elsewhere, who would quarrel with this argument.

The author's view is that safety is always paramount, and effectiveness studies other than those required for risk/benefit assessment are a luxury for administrations with effort to spare. So, in considering just how close the approaches of the various administrations might be brought in the future, I suggest that the following factors be borne in mind:

- the number of genuinely novel devices is small
- the objective of the approval process should be limited to deciding whether the device can be allowed entry to the market
- the clinical evaluation should not attempt comparisons with alternative devices or procedures—these should be left for post-market studies
- the clinical investigation should be used to establish the key engineering parameters to be used in the approval of similar devices
- the limitations of pre-market evaluations should be accepted.

The last of these factors is particularly important. Authors already quoted in this chapter have pointed out the limitations of pre-market clinical trials, and experience of late problems with approved devices, such as the Shiley heart valve, silicone breast implants and several types of artificial hip joints, have brought home that trials of any practicable length cannot expose such problems. The public health therefore demands other measures to reduce the impact of late problems and these are discussed in the following chapter.

Chapter 9

Key factors—post-market controls

Introduction

One of the most significant developments in medical device regulation has been the emergence of an appreciation that pre-market approval processes cannot eliminate all possibility of problems. Some of the problems which have gained world-wide attention in recent years, such as polyurethane pacemaker leads (Stokes 1998), the Bjork-Shiley heart valve (Lindblom *et al.* 1989), silicone gel breast implants (Snyder 1997) and several kinds of artificial hip joint, have become apparent many years after implantation.

The fact that the examples quoted are all implanted devices should not be taken to imply that only long-term implants suffer problems after going through an approval process. Many other types of medical device have experienced problems which have had, in some cases, serious consequences (e.g. Globe 86, EFOMP 91) but rapid corrective action is generally possible in the case of external devices without the arousal of public concern. The difficulties associated with corrective action in the case of implants calls for special attention to such devices in both the pre-market and post-market stages. Pre-market approval methods have been given most attention in the past, and key features of pre-market approval systems are discussed in Chapters 6, 7 and 8. This chapter will examine the current and future approaches to post-market controls.

Three distinct (but not necessarily independent) techniques are in use today; adverse event reporting, post-market surveillance and implant registration/tracking.

Adverse event reporting

Adverse event reporting has a long history. In the UK the principle that defects in medical devices should be reported to the Scientific and Technical

Branch of the Department of Health and Social Security was well established in the 1960s. Instructions to the National Health Service about the reporting of such defects were repeated in 1974 (UK 74) and 1983 (UK 83). This was an entirely voluntary system aimed at the users of medical equipment found in NHS hospitals. The operation of the system is described in Higson (1983) and in Health Equipment Information No. 98 (UK 82a).

A similar voluntary user reporting system also existed in France, but Anhoury (1994) comments that the 'alert forms' issued by the Ministry of Health were largely ignored by manufacturers and users—only 23 alert forms were received by the Ministry in 1989. This may be compared with 969 reports received by the UK DHSS in 1982 (Higson 1983), rising to more than 3500 in 1994 (UK 95) and to over 6000 in 1998 (UK 99).

In Germany, paragraphs 11.3 and 15 of the MedGV (Germany 85) placed an obligation on users to report adverse events but Wolf (1994) claims that this obligation was not implemented in practice. Likewise, a similar obligation existed under Spanish law (Spain 86), bearing on manufacturers, importers and users.

A voluntary reporting system was established in the United States by ECRI, a US-based non-profit agency, in 1971 (Nobel 1994) and was followed by a voluntary system set up by the FDA in 1975. A voluntary system exists also in Australia where 681 reports were received in 1995 (Australia 96b).

A major change in the basis of adverse event reporting occurred in 1984 with the issuing of medical device reporting regulations in the USA which required manufacturers to report to the FDA any adverse event in which a medical device might have been involved. To the author's knowledge, this was the first time that such reporting had been made mandatory.

The Medical Device Reporting Regulations were published in September 1984 (USA 84) and introduced the following requirement:

'FDA is requiring a device manufacturer or importer to report to FDA whenever the manufacturer or importer receives or otherwise becomes aware of information that reasonably suggests that one of its marketed devices:

(1) may have caused or contributed to a death or serious injury, or
(2) has malfunctioned and that the device or any other device marketed by the manufacturer or importer would be likely to cause or contribute to a death or serious injury if the malfunction were to recur.

In addition, a device importer is required to establish and maintain a complaint file and to permit any authorized FDA employee at all reasonable times to have access to and to copy and verify the records contained in this file.'

An incident as described in indent (1) is to be reported by telephone within 5 days, and in writing within 15 days. A malfunction is to be reported in writing within 15 days.

The Safe Medical Devices Act of 1990 (USA 90a) extended reporting requirements similar to those quoted above to distributors and to device user facilities, and introduced annual certification of reports submitted.

The Medical Device Amendments of 1992 (USA 92) added a requirement for manufacturers, distributors or importers to report any 'other significant adverse device experiences as determined by the Secretary (of Health and Human Services) to be reported' and added a description of a serious injury to mean an injury that

'(A) is life threatening,
(B)   results in permanent impairment of a body function or permanent damage to a body structure, or
(C)   necessitates medical or surgical intervention to preclude permanent impairment of a body function or permanent damage to a body structure.'

The latest changes in the reporting requirements are contained in Section 213 of the Food and Drug Modernization Act of 1997 (USA 97a). This amends Section 519 of the Food Drug and Cosmetic Act to remove the reporting requirement from distributors and directs FDA to promulgate a new regulation requiring distributors to keep records of adverse events and make them available to FDA on request. It also revoked the requirement for manufacturers, distributors and importers to certify annually the number of adverse events submitted during the year. These, and other minor changes, were brought into effect by a Direct Final Rule issued in May 1998 and replaced by a Final Rule issued in January 2000 (USA 00a).

Reporting under the Medical Devices Directive

The MDD (EC 93a) follows the example set by the Medical Device Amendments of 1976 by making reporting by manufacturers of adverse incidents mandatory. This is achieved by including in all the conformity assessment Annexes to the Directive (Annexes II–VII) an obligation for the manufacturer to give an undertaking to notify the Competent Authority of all incidents that involve:

'(i) any malfunction or deterioration in the characteristics and/or performance of a device, as well as any inadequacy in the labelling or the instructions for use which might lead to or might have led to the death of a patient or user or to a serious deterioration in their state of health;
(ii) any technical or medical reason connected with the characteristics or performance of a device for the reasons referred to in sub-paragraph (i) above leading to a systematic recall of devices of the same type by the manufacturer.'

Furthermore, Article 10 of the MDD requires Member States to ensure that all such incidents brought to their attention are recorded and evaluated centrally. It specifically permits Member States to require users to report adverse incidents and several have done so. It also requires Member States to inform the Commission and the other Member States of any incidents which, after investigation, have resulted in some form of corrective action—irrespective of whether such reports may have been made as a result of action taken under the Safeguard Clause (see Chapter 3).

Guidance on the application of these requirements was prepared by the European Commission, in conjunction with the Competent Authorities and the leading European Trade Associations, and issued for the first time in 1993 and in revised form in 1998 (EC 98h). This document gives useful guidance on the types of incident to be reported, reporting procedure, time scales, incident investigations and cooperation between Competent Authorities.

A 'serious deterioration in state of health' is defined as:

'can include:
— life-threatening illness or injury;
— permanent impairment of a body function or permanent damage to a body structure;
— a condition necessitating medical or surgical intervention to prevent permanent impairment of a body function or permanent damage to a body structure.'

This is virtually identical with the US definition quoted above.

Detailed guidance, with examples, is given on 'near incidents', 'inadequacies in the labelling or instructions for use', the Competent Authority to which the report is to be made, and details to be included in the initial report.

Incidents are to be reported within 10 days and near incidents within 30 days. Examples of reporting forms are included in the guidance but several Member States have issued their own forms.

Manufacturers are normally expected to perform the investigations of incidents, monitored by the Competent Authority which may need to notify other CAs in the case of a serious risk to patients elsewhere in the Community or where coordination is needed because of multiple reports. The 1998–99 Annual Report of the UK Medical Devices Agency (UK 99) notes that the MDA received 454 adverse event reports from manufacturers under the Medical Devices Regulations (UK 94b), of which 35 were communicated to other Competent Authorities, and they received 57 notifications from other Member States. (These numbers may be compared with 6125 reports received by the MDA from the NHS under the voluntary defect reporting system.)

The Commission's 'Guidelines on a Medical Device Vigilance System', supplemented by additional guidance issued by the UK Medical Devices Agency (UK 98a, UK 98b, UK 98c) constitutes probably the most complete guidance available on adverse incidents and their reporting. These documents are now being revised to bring them into line with the recommendations of the Global Harmonization Task Force (see below).

Adverse event reporting in Japan

Japan, also, has similar reporting requirements in Article 62-2 of the Enforcement Regulations (Japan 95b) of the Pharmaceutical Affairs Law 1994:

> 'When the manufacturers or importers of medical devices, or the recipients of approval of foreign manufacture learn of any of the following information related to the medical devices which they manufacture or import, or for which they have obtained approval for manufacture, they shall so report to the MHW minister within the period specified below...
>
> (1) Within 15 days if the information is related to the following.
>   A. Unexpectable death, disablement or other related cases suspected to be caused by deficiencies in the medical devices being used in Japan. But, this is also applicable to those devices being used abroad, if they are similar in terms of shape, construction, raw materials, usage, indications, effect, properties, etc....
>   B. Deficiencies of medical devices, which might lead to the cases mentioned in A.
> (2) Within 30 days if the information is related to the following, but excluding the preceding item.
>   A. Death, disablement or other related cases which are suspected to be caused by deficiencies in the medical devices, and which are almost incurable or found serious by the physician or dentist in charge.
>   B. Disorders which are suspected to be caused by deficiencies in the medical device and which are found not insignificant by the physician or dentist in charge...
>   C. Deficiencies of medical devices, which might lead to the cases mentioned in A or B.
>   D. Research reports about the following: possible deficiencies in medical devices...; remarkable changes in the trend of incidence...; or the lack of the approved indications, effects or properties.'

Reports are to be made on forms issued by the Ministry (Japan 92b) and submitted to the Safety Division of the Pharmaceutical Affairs Bureau.

Harmonization of adverse event reporting

A study carried out by Study Group 2 of the Global Harmonization Task Force (see Chapter 10) (GHTF 00a) showed that the adverse event reporting requirements in the four administrations studied (Australia, EC, Japan, USA) were very similar. As a result, SG2 was given the tasks of fully harmonizing the requirements and establishing a system for exchange of information about adverse events as priority tasks.

Study Group 2 has now prepared an agreed set of rules defining events which are to be reported, together with guidance on the use of the rules (see Appendix 4). A minimum data set for adverse event reports has also been prepared and further work is expected to result in an optimum data set. No agreement has yet been reached on the reporting time; however, the work done so far should make it possible for manufacturers to prepare one report suitable for all authorities world-wide and submit it in all countries where the device in question is sold. SG 2 has also begun work on the definition of a system for sharing reports among national authorities.

Progress of the work of Study Group 2 can be monitored on the Internet at: http://www.ghtf.org

Post-market surveillance

In addition to the long-established practice of the reporting of adverse events, the systematic follow-up of devices in use has been introduced in recent years as a means both of accelerating the awareness of problems and of facilitating action to mitigate the effects of such problems on patients.

Requirements in the United States

Again, the first approach to post-market surveillance was made in the United States. The Safe Medical Devices Act of 1990 (USA 90a) introduced a section 522 to the Federal Food, Drug and Cosmetic Act (FDCA) authorizing the FDA to conduct post-market surveillance for devices that are permanent implants, the failure of which may cause adverse health consequences or death; or that are intended for use in supporting or sustaining human life; or that potentially present a serious risk to human health. Furthermore, a manufacturer could be required to conduct post-market surveillance of any device if the FDA determined that such surveillance was necessary to protect the public health or to provide safety or effectiveness for the device. Manufacturers were directed to provide protocols for the required surveillance within 30 days of the initial marketing of the product. The 1992 Medical Device Amendments (USA 92) made violation of the post-market surveillance requirements a violation of the FDCA and changed the time for submitting a protocol to 30 days after the manufacturer was informed of the requirement.

These provisions placed heavy burdens on manufacturers and were appreciably relaxed by the FDA Modernization Act of 1997 (USA 97a) so that post-market surveillance is now required only when the FDA issues an order; and such orders may only be issued in respect of

'a Class II or Class III device the failure of which would be reasonably likely to have adverse health consequences or which is intended to be—
(1) implanted in the human body for more than one year, or
(2) a life sustaining or life supporting device used outside a device user facility.'

The method(s) of conducting post-market surveillance is not prescribed, but the manufacturer has to submit a plan within 30 days of receiving an order from the FDA. The FDA 'within 60 days of the receipt of such plan, shall determine if the person designated to conduct the surveillance has appropriate qualifications and experience to undertake such surveillance and if the plan will result in the collection of useful data that can reveal unforeseen adverse events or other information necessary to protect the public health.'

Guidance on post-market surveillance was issued in November 1998 (USA 98e). As a rationale for post-market surveillance, it states 'Pre-market testing cannot address all device-related concerns. While post-market surveillance will not be used in lieu of adequate pre-market testing, post-market surveillance can serve to complement pre-market data. Certain issues that arise during pre-market evaluation of a device may be more appropriately addressed through data collection in the post-market period rather than prior to approval/clearance for marketing.'

Requirements in Europe

The EC authorities were also concerned to introduce an element of post-market surveillance into the Medical Devices Directives. The early drafts of the AIMDD included the term 'post-market surveillance' and during the attempt to provide a workable definition it was decided to incorporate the definition directly in the text as 'a systematic procedure to review experience gained from devices in the post-production phase'.

Unfortunately, this phrase does not contain adequate guidance on the scale and scope of the review procedure and there is some doubt about just what is required of manufacturers of various types of device. The manufacturer must make a declaration to institute and keep up-to-date such a review system as part of the application for approval of the quality system under Annexes II, V and VI, as part of the EC Verification under Annex IV, and as a constituent of a Declaration of Conformity for Class I products under Annex VII. Thus, apart from the Class I products, it is for the Notified Bodies to decide on the acceptability of the proposed system. The Notified Bodies Group has attempted to address the problem by producing a recommendation: NB-MED 2.12 (NB 98b).

This paper takes the view that 'PMS systems are based on information received from the field (e.g. complaint monitoring, feedback from sales representatives, reports from regulatory authorities, literature reviews, service/repair information) and its analysis as described and referred to in EN 46001/2 (CEN 97a, 97b) clause 4.14.' This clause states 'The supplier shall establish and maintain a documented feedback system to provide early warning of quality problems and for input into the corrective action system.... All feedback information, including reported customer complaints and returned product, shall be documented, investigated interpreted, collated and communicated in accordance with defined procedures by a designated person.' The Notified Bodies are, therefore, relying on their inspections of manufacturers' quality systems to satisfy themselves that this requirement of the Directives is met. The 'passive' approach described in the quality system standards is acceptable for the majority of medical devices, but could be seen as somewhat limited for genuinely novel Class III devices and many implants for which a more active search for possible reactions and side effects might be more appropriate. A discussion of methods of post-market surveillance is given below.

Another form of post-market surveillance is that carried out by the Competent Authorities to discharge their obligations under Article 2 of the Directives, i.e. 'Member States shall take all necessary steps to ensure that devices may be placed on the market and put into service only if they do not compromise the safety and health of patients...'. At least the UK Competent Authority (the Medical Devices Agency) interprets this Article as requiring them to check devices on the market for compliance with the Directives. In a letter to Trade Associations early in 1997 (UK 97a) the MDA announced its intention to 'take a pro-active role in ensuring compliance with the (Medical Device) Regulations' and its conclusion that it should 'concentrate our efforts on Class I and custom made devices where, unless the device is sterile or has a measuring function, there is no Notified Body involvement, plus those companies who assemble medical devices or sterilize them...'.

In its Annual Report for 1998–99 (UK 99), the MDA states that 55 inspections of manufacturers of such products were carried out during the year as well as the investigation of 195 cases of alleged non-conformity reported by third parties. The MDA concludes 'Where non-conformities were found, most were of a minor and technical nature.'

The author is unaware of any other European Competent Authority which has mounted a similar programme. Few Competent Authorities have sufficient staff to do so and none has staff numbers approaching those of the MDA (137) (UK 99).

Requirements in Japan

The Pharmaceutical Affairs law and its associated Ordinances and Enforcement Regulations do not contain any specific post-market surveillance requirements but the Guide to Medical Device Registration in Japan (Japan 94c) gives extensive guidance on what is expected of both manufacturers and others involved in the provision and use of medical devices. In Chapter 11, Section 3(1), it states 'The manufacturers should establish an independent section for collection and distribution of efficacy and safety data in a post-marketing surveillance. Any data should be collected directly not only from physicians and dentists but from related literature, and correctly arranged and evaluated. The results of the evaluation and new findings and important information in the package insert should be rapidly communicated to medical service personnel. Manufacturers should promote the post-marketing surveillance by making their own bylaws.' This excellent guidance is supplemented later in the same chapter by detailed guidance on the organization manufacturers should adopt to enable post-market surveillance to be carried out efficiently.

Uncertainties for manufacturers

It remains the case that manufacturers are unsure about the most appropriate techniques of post-market surveillance to apply to different types of device, and about how extensive their surveillance programme should be. Several possibilities are apparent:

–   The pooling of such reports at a central point and their systematic analysis in search of recurring problems or faults is an obvious first approach which satisfies the requirement of the MDD for many, probably most, devices. The inclusion of information arising from other sources such as user complaints, feedback from sales and service personnel, reports in the literature, etc. can improve the process.
–   For equipment in receipt of preventive maintenance, regular review of service reports or equipment logs, and the pooling of the information, can reveal systematic problems.
–   For any kind of failed device, investigation of the mode of failure and association with other failure reports, can give valuable insights. Witkin (1998c) points out that such investigations are particularly useful for implanted devices.
–   For the more serious devices, particularly those subjected to clinical investigation as part of the approval process, long term clinical follow-up at a few centres is essential if failures or side effects are to be identified at the earliest stages. Witkin (1998c) gives useful guidance on the conduct of such clinical follow-up.
–   For implanted devices, implant/explant registries can be a valuable source of information about problems as they begin to appear (see below).

The FDA Guidance on Criteria and Approaches for Post-market Surveillance (USA 98e) lists very similar 'approaches for post-market surveillance':

- Detailed review of complaint history and scientific literature
- Non-clinical testing of the device
- Telephone or mail follow-up of a defined patient sample
- Use of existing secondary data sets, such as Medicare data
- Use of registries, such as the Society for Angiography and Interventional Cardiology (SAIC) stent registry, or internal registries or tracking systems
- Case-control study of patients implanted with or using devices
- Consecutive enrolment studies
- Cross-sectional studies (multiple cohorts)
- Randomized controlled trials.

Study Group 2 of the Global Harmonization Task Force has announced its intention to produce guidance on post-market surveillance when it has finished its present work on adverse event reporting. Guidance from such an authoritative source would be advantageous.

Patient registries/tracking

Implanted devices present particular problems of post-market control because of the mobility of the patients and the patients' general lack of awareness of the specific type of device that has been implanted. One of the first approaches that was made to the solution of these problems was the establishment of implant data banks.

These require the notification of the implantation and explantation of designated devices to a central point or registry. The first of these was probably the Pacemaker Data Bank operated by the National Heart Hospital on behalf of the Department of Health and Social Security (Shenolikar and Worrol 1982) and which began in 1978. This was built on the back of schemes already operated by individual pacemaker manufacturers and took advantage of the standardized registration card introduced by the International Association of Pacemaker Manufacturers (IAPM) (now a sector of EUCOMED) (Rickards 1985). This scheme was entirely voluntary and depended on the cooperation of implanting surgeons which was given to a high degree.

The prime objective was to collect explant information as well as implant data so that premature failures could be recognized early. Although explant information is difficult to gather (as explantation or patient death may occur anywhere—not necessarily in the implanting hospital—and as death may result from causes not related to the implant, the need to notify the registry of the patient details may not be known) early warning of problems from the analysis of registry data was claimed on several occasions. The pacemaker scheme was regarded as so successful that in 1986, following the emergence of problems with the Bjork-Shiley heart valve (Lindblom *et al.* 1989), a similar registry was established for heart valves at the Hammersmith Hospital (Taylor *et al.* 1992). It has recently been reported (Cutts 1999) that a National Hip Implant Registry is under active consideration.

Although these registries were set up and financed by the Department of Health and Social Security (as it was then) they had to be run by hospital departments both to engage the cooperation of implanting clinicians and to overcome issues of patient confidentiality. The information passed to the DHSS was statistical information relevant to the performance of the different types of pacemaker and heart valve implanted in the UK and all links with individual patients were removed. Tracking to individual patients when a problem was identified could only be done via the implanting clinician and was virtually impossible when they had lost touch.

It should be noted that, although patient registries are recognized as an effective method of post-market surveillance (and are specifically mentioned by the Notified Bodies Group (NB 98b) as satisfying the requirement of the Active Implantable Medical Devices Directive), they are not legally enforced, to the author's knowledge, in any administration at this time, whereas tracking of devices to individual patients is a requirement of the legislation in Japan and USA.

Implanted devices pose special problems when defects in them are discovered. These arise from the need to locate the devices, and the hazards associated with the surgery necessary if explantation is indicated. Tracking of implantable devices, i.e. linking of the devices with identified patients and monitoring the location of the patients, was introduced in the USA by the Safe Medical Devices Act of 1990 (USA 90a). This required a manufacturer of '(A) a permanently implantable device, or (B) a life sustaining or life supporting device used outside a device user facility, or ... any other device which the Secretary may designate' to adopt a method of device tracking. The FDA Modernization Act of 1997 (USA 97a) limited this authority by revising Section 519(e) of the FD&C Act to read:

'The Secretary may by order require a manufacturer to adopt a method of tracking a class II or class III device—
   (A) the failure of which would be reasonably likely to have serious adverse health consequences; or
   (B) which is—
       (i) intended to be implanted in the human body for more than one year, or
       (ii) a life sustaining or life supporting device used outside a device user facility.'

Patients have the right to refuse to release any identifying information for the purpose of tracking.

Guidance on tracking was published by the FDA in February 1999 and replaced in January 2000. It was followed by the publication of a Proposed

Rule on tracking (USA 00c). The guidance is available on the website: http://www.fda.gov/cdrh/modact/tracking.pdf

At this time the following devices must be tracked by the manufacturers:

Implants
- Temporomandibular joint (TMJ) prosthesis
- Glenoid fossa prosthesis
- Mandibular condyle prosthesis
- Implantable pacemaker pulse generator
- Cardiovascular permanent implantable pacemaker electrode
- Replacement heart valve (mechanical only)
- Automatic implantable cardioverter/defibrillator
- Implanted cerebellar stimulator
- Implanted diaphragmatic/phrenic nerve stimulator
- Implantable infusion pumps
- Dura mater
- Abdominal aortic aneurysm stent grafts

Non-implants (used outside a device user facility)
- Breathing frequency monitors
- Continuous ventilators
- Ventricular bypass (assist) device
- DC-defibrillators and paddles
- Infusion pumps (electromechanical only).

Tracking is, in fact, difficult and costly for manufacturers to do and it is not known how accurately it is being done. Nevertheless, this requirement has been introduced into legislation in other countries, probably as a result of the problems with the Bjork-Shiley heart valve (Lindblom *et al.* 1989). Tracking of five types of implantable device was introduced in Japan by the 1994 revision of the Pharmaceutical Affairs Law (see Chapter 4).

As has been noted, tracking is not only difficult for manufacturers to carry out but it demands a high level of cooperation from patients. The operation of patient registries or data banks is much simpler. Registries of this kind are not intended to allow the location of patients with a specific type of implant but they do make it possible to link a patient to the implanting hospital with which patients commonly keep in contact and, as they are much simpler to operate than tracking schemes, they are probably much more cost effective.

It is worth noting that the introduction of mandatory reporting of adverse events linked to medical devices has resulted in the identification of problems with implanted devices such as heart valves much more rapidly than was previously possible and the UK Medical Devices Agency has decided that it is no longer necessary for them to fund the Heart Valve Registry in order to gather this kind of information. However, the medical information also available from the Registry (see, for example, Ratnatunga *et al.* 1998) is of such value that the Registry is continuing with funding from the Policy Divisions of the Department of Health (Tinkler 1998). The value of the long-term outcome information available from registries in health technology assessment is emphasized also by Rickards and Cunningham (1999).

Commentary

The introduction of post-market surveillance requirements in many regulations from 1990 onwards is a clear sign that the limitations of pre-market approval mechanisms are now recognized. This is perhaps most obviously the case for orthopaedic implants for which the conditions of use are practically impossible to reproduce in the laboratory. Furthermore, device failures attributed to inadequate quality control, such as those in the case of the Bjork-Shiley heart valve, occur well after the approval process and may exist for a long period before detection.

Rapid detection of failures, adverse reactions or side effects therefore has to be regarded as an essential feature of protection of the public health. The routine reporting of incidents as they occur is no longer accepted as sufficient—it is the 'systematic procedure to review experience gained from devices in the post-production phase' (as required by the EC medical device Directives) which offers the most rapid appreciation that a problem exists.

The value of early detection and/or awareness of problems with a medical device is obviously based on the saving of life or injury from the same cause by rapid remedial action. Even greater value can be realized if modifications to eliminate the problem can be introduced on a general scale—often by means of a new or amended product standard (as discussed in Chapter 7). The rapid exchange of information between regulatory authorities, as required by the EC medical device Directives and as proposed by the Global Harmonization Task Force, would clearly increase value in both senses mentioned here.

The methodology of post-market surveillance is not well understood and international agreement on methods appropriate to different kinds of device is needed. Guidance from the Global Harmonization Task Force, supplementing its guidance on adverse event reporting, is awaited and is expected to be definitive.

Defects in non-implanted devices are generally easily dealt with by taking all devices of the defective type out of service and, wherever possible, making appropriate modifications or restricting the conditions of use. Defective implants are much more difficult to deal with because of the nature of the surgery involved, but also because of the difficulty of locating the devices. For this reason tracking requirements are imposed by several administrations. It seems questionable to impose requirements on manufacturers that they may be unable to fulfil—certainly they need the cooperation of patients and hospitals if they are to meet the requirements.

It is now clear that a uniform definition of a reportable event has been reached and that information exchange mechanisms will be in place between the major administrations in the foreseeable future. These will undoubtedly promote more rapid appreciation of problems with devices. Improvements in methods of post-market surveillance will be a further step towards early detection of problems and possible future technological developments may solve or ease the difficulty of locating patients with problem implants.

Chapter 10

Proposals and prospects for a global regulatory system for medical devices

This book has described the emergence of regulations for medical devices during the 1960s and 1970s, the transition (still in progress) from the treatment of medical devices as a particular type of pharmaceutical to an engineering approach, and the identification of features regarded as 'key' in any system for assuring the safety and satisfaction of medical devices. The general acceptance of these key features and their prominence in the most recent regulations are evidence of trends leading regulatory authorities towards a common approach which offers prospects of a global regulatory system at some point in the future (Higson 1995). This process has been led in the last few years by the Global Harmonization Task Force (Higson 1996), supported by ISO Technical Committee 210.

The Global Harmonization Task Force

The Global Harmonization Task Force (GHTF) was formed during the Third Global Medical Device Conference held in Nice in September 1992. It followed from the recognition that the decision by the FDA to revise the US Good Manufacturing Practice Regulation and bring it into line with ISO 9000 standards provided an opportunity—possibly never to recur—to harmonize quality assurance requirements for medical devices between the US and the EC. James Benson, then Director of the Center for Devices and Radiological Health, was persuaded to write to John Mogg of the European Commission outlining this opportunity and seeking the co-operation of the Europeans. The Commission gave a positive response and produced a paper (unpublished, reproduced at Appendix 1) presented to participants in the Nice meeting which resulted in a commitment from industry and regulatory authorities in the EC, Japan, USA and Canada to work together to reduce the differences in their regulatory systems. The Chair and Secretariat were taken by the European Commission and the first task chosen was the harmonization of the quality systems used in the four participating markets (subsequently expanded to include Australia). The progress and success of this activity is described in Chapter 6.

The work of the GHTF was seen by the European Commission as quite distinct from the negotiation of mutual recognition agreements (MRAs) between the EC and other countries as mandated in the Council Resolution of 21 December 1989 (EC 89a). Whereas MRAs are driven by political and commercial considerations, the GHTF was to be informal and primarily technical in nature.

The GHTF carries out its work in Study Groups (SGs) which meet independently and conduct their own agendas subject to the overall remit given to them by the GHTF. There are currently four Study Groups:

SG1   Comparison of regulations (convenor: EC)
SG2   Harmonization of rules for adverse event reporting and post-market surveillance: global information exchange (convenor: USA)
SG3   Guidance documents on quality systems and associated aspects (convenor: USA)
SG4   Harmonization of quality system auditing practices; training and qualification of auditors (convenor: EC).

The work on quality system guidance and auditing satisfied the original remit of the GHTF and dominated the first two years of work. A guidance document for manufacturers on the establishment of quality systems was produced in 1994, given wide circulation among the industry constituents and passed to ISO where it became the starting point for ISO 14969 (ISO 99a). Constructive comments were offered on the draft US and Japanese GMP Regulations and undoubtedly were influential.

SG1 produced a brief study of the regulations in the four participating administrations which was presented at the second GHTF meeting, and this led to an expansion of the activities of the GHTF when SG1 was given revised terms of reference in 1994:

'To examine medical device regulatory systems in use in the major trading countries or regions, and to:
–   identify features of those systems which have a common basis but different application;
–   identify features peculiar to individual systems which may present obstacles to uniform regulation;
–   make proposals to the GHTF for harmonization activities relating to these features, *and*
–   suggest priorities.'

As a result of this programme, SG1 identified and proposed to the GHTF the work on adverse effect reporting which was then allocated to SG2, commented on the Canadian Regulations (Canada 98) at draft stage, and began developing fundamental papers on 'Essential Principles', Reference to Standards, and Classification. Work on universal dossiers for approval applications also shows promise of receiving general acceptance and the views of the GHTF on emerging regulations is being perceived as valuable. The range of topics considered by the GHTF for possible activity continues to expand—the 2000 meeting held sessions on nomenclature, the use of PVC in medical devices and re-use of single-use devices. As the influence of the GHTF increases, attendance at its (generally annual) meetings extends well beyond the five members, and meetings of the Asian and South American groups, who seek affiliation to the GHTF, usually take place at the same time.

Outputs of the GHTF

The following documents have been posted on the internet (www.ghtf.org), or are in the final stages:

Guidance on Quality Systems for the Design and Manufacture of Medical Devices (GHTF.SG3-N99-8)	Issued August 1994, Posted 30 October 2000 (GHTF 94)
Design Control Guidance for Medical Device Manufacturers (GHTF.SG3-N99-9)	11 September 1997 Posted 30 October 2000
Process Validation Guidance for Medical Device Manufacturers (GHTF.SG3-N99-10)	20 February 2000 Posted 30 October 2000
Guidelines for Regulatory Auditing of Quality Systems of Medical Device Manufacturers (GHTF.SG4(99)28)	30 December 1999 Posted 30 October 2000
Essential Principles of Safety and Performance of Medical Devices (GHTF.SG1.NO20)	Posted 28 December 1999 (Appendix 2)
Role of Standards in the Assessment of Medical Devices (GHTF.SG1.NO12)	15 March 2000 Posted 23 October 2000 (Appendix 3)
Labelling for Medical Devices (GHTF.SG1.NO09) (GHTF 00b)	15 March 2000 Posted 23 October 2000
Medical Device Classification (GHTF.SG1.NO15)	Draft 10 January 2001 (GHTF 01)
Summary Technical File for Pre-market Documentation of Conformity with Requirements for Medical Devices (GHTF.SG1.NO11) (GHTF 99)	Draft 28 December 1999
Adverse Event Reporting Guidance for the Medical Devices Manufacturer or its Authorized Representative (GHTF.SG2.N21)	29 June 1999 Posted 15 October 2000 (Appendix 4)

This is not an exhaustive list of GHTF documents but it includes those regarded by the author as most significant.

These documents (and others still in progress) constitute the building blocks of a rational system for assuring the safety and performance of medical devices. Such a system would apply measures for the compliance of devices with 'essential principles' and would need a body of standards to complement them. Many standards have been and are being produced in the IEC and ISO but a key role in ensuring the coherence and completeness of the portfolio of standards is played by ISO Technical Committee 210 *Quality management and corresponding general aspects for medical devices.*

ISO Technical Committee 210

At its meeting in September 1993 the ISO Technical Board agreed to recommend to the ISO Council

'the establishment of a new technical committee with the following provisional title:

General aspects for health care products

its responsibilities to include the preparation of horizontal standards for the application of the ISO series of standards to health care products as well as for fundamental requirements for health care products.'

The first meeting of TC 210 was unable to accept this mandate as presented and modified it to the following title and scope, which was then agreed by the Technical Board:

'*Quality management and corresponding general aspects for medical devices*

The standardization of requirements and guidance in the field of quality management for medical devices; the development of standards on general aspects stemming from the application of quality principles to medical devices, where these are not covered by the scope of another technical committee.'

Despite the tortuous wording of the scope, TC 210 was left with considerable freedom to decide its own work programme which was divided between four Working Groups:

WG1   Application of quality systems to medical devices
WG2   General aspects stemming from the application of quality principles to medical devices
WG3   Symbols, definitions and nomenclature for medical devices
WG4   Application of risk management to medical devices.

A close collaboration with the Global Harmonization Task Force, other ISO committees, IEC, CEN and CENELEC has evolved and the following documents have been, or are being, produced:

ISO 13485:1997	Quality systems—Medical devices—Particular requirements for the application of ISO 9001. (ISO 96a—corresponds to EN 46001 (CEN 97a)) (new version to apply to ISO 9000:2000 is in preparation).
ISO 13488:1997	Quality systems—Medical devices—Particular requirements for the application of ISO 9002 (ISO 96b—corresponds to EN 46002 (CEN 97b)) (new version to apply to ISO 9000:2000 is in preparation).
ISO 14969:1999	Quality systems—medical devices—Guidance on the application of ISO 13485 and ISO 13488 (ISO 99a—corresponds to Guidance from the GHTF (GHTF 94)).
ISO 14971-1:1998	Medical devices—Risk management—Part 1: Application of risk analysis to medical devices (corresponds to EN1441 (CEN 97c)).
ISO 14971:2000	Medical devices—Risk management—Application of risk management to medical devices. (ISO 00b—will replace ISO 14971-1 and EN 1441).
ISO 15223:2000	Medical devices—Symbols to be used with medical device labels, labelling and information to be supplied (supersedes ISO/TR 15223:1998) (ISO 00f—corresponds to EN 980 (CEN 97d)).
ISO 15225:2000	Nomenclature—Specification for a nomenclature system for medical devices for the purpose of regulatory data exchange (issued identically in CEN as EN ISO15225).
ISO/CD 16142	Medical devices—Guide to the selection of standards in support of recognized essential principles for safety and performance of medical devices (ISO 99b).
ISO/IEC Guide 63	Guide to the development and inclusion of safety aspects in international standards for medical devices (ISO/IEC 99a).

It can readily be seen that the work of ISO/TC 210 is closely associated with that of the Global Harmonization Task Force and with the European standards programme being carried out in response to mandates from the European Commission. A Memorandum of Understanding clarifying the relationship between the GHTF and ISO/TC 210 was adopted by ISO/TC 210 at its meeting in April 1999 and by the GHTF at its meeting in

June 1999. Under this MoU, the GHTF has liaison status with TC 210 and TC 210 has the right to participate in GHTF Task Force and Study Group meetings.

Trends in medical device regulation

The activities of the GHTF and ISO/TC 210, and other events described throughout this book, seem to provide confirmation of a number of trends observed by the author:

- A move from a 'pharmaceutical approach' characterized by a general requirement for safety and effectiveness to an 'engineering approach' characterized by 'essential principles' deemed to be satisfied by compliance with standards;
- The expansion of conformity assessment from dossier examination for substantial equivalence or PMA to such methods as self certification, third-party certification, type testing and quality systems;
- A move from post-market checks on compliance of manufacturing operations with GMP to the use of pre-market compliance with quality system requirements based on ISO 9001/ISO 13485 (ISO 94a/ISO 96a) as a form of conformity assessment;
- A recognition of the limitations of pre-market safety assessments and increased reliance on adverse event reporting and post-market surveillance;
- A cautious introduction of third party assessment of both products and quality systems as a replacement for total reliance on assessment by the regulatory authorities.

The continuation of these trends, coupled with an increase in the number of international product standards complying with ISO/IEC Guide 63 (ISO/IEC 99a), could and should lead to an internationally-accepted system of medical device regulation with the basic principles described in the following Section.

A global regulatory system

Outline of a global system

A global regulatory system for medical devices is likely to have the following features:

1. All medical devices (as defined in the MDD) must conform to the essential principles of safety and performance as promulgated by the Global Harmonization Task Force (see Appendix 2).

2. Compliance with 'recognized standards' will be deemed to satisfy the essential principles. The recognized standards will be identified by each regulatory authority but will include a core group of international standards generally accepted. This core group will be based on ISO/CD 16142 (ISO 99b) which may be superseded by the output of a possible subsidiary Study Group of the GHTF, which could be established to identify standards suitable for recognition according to the principles set out in GHTF.SG1.NO12 (see Appendix 3).
3. Devices will be placed in one of four classes according to the final version of the GHTF classification system (GHTF 01).
4. The assessment of conformity with the essential principles will be progressive in severity according to the class of the device, generally as follows:

  Class A   A manufacturer's declaration of conformity with the essential principles, coupled with a declaration of compliance with ISO 9003 (ISO 94c) or ISO 13488 (ISO 96b) (see Note 1).

  Class B   A manufacturer's declaration of conformity with the essential principles, coupled with certification by a recognized body (see Note 2) of compliance with ISO 13488.

  Class C   A manufacturer's declaration of conformity with the essential principles, coupled with certification by a recognized body of compliance with ISO 13485 (ISO 96a) (see Note 3).

  Class D   As Class C, but with a review by a recognized body of the evidence showing compliance with the essential principles and that a clinical evaluation (see Chapter 8) has been carried out.

EXPLANATORY NOTES

  (1) The system proposed here departs from both the EC Medical Device Directive and the US Quality System Regulation in making compliance with a quality system of some kind a requirement for all classes, but in grading the quality system according to class. ISO 9003, which requires only a final inspection and test of the product, is suitable for many Class A devices such as ostomy devices, orthoses, and reusable surgical instruments, but where sterilization or any other special manufacturing process is involved, this quality standard is inadequate and conformity with ISO 13488 is necessary. This difference results in an effective subdivision of Class A into Classes Ai and Aii.

  (2) A 'recognized body' may be the regulatory authority itself or a non-governmental institution. Non-governmental bodies must be appointed by, responsible to, and monitored by the regulatory authority and must comply with applicable ISO/IEC Guides.

(3) ISO 13485 (ISO 96a) includes design controls which require the manufacturer to verify[1] and validate[2] the design of the medical device. As noted above (Chapter 8) the validation of medical devices will generally include clinical evaluation and may include clinical investigation.

5. Devices which have been through the appropriate conformity assessment procedure will be marked with a globally-recognized mark.
   N.B.
   A mark of this kind is needed to facilitate the acceptance of conforming devices but the use of this mark will have to be protected. This will demand agreement between participating administrations to enforce national sanctions on persons misusing the mark.

6. Devices and their packaging will be labelled in accordance with the recommendations of the GHTF (GHTF 00b), using symbols according to ISO 15223 (ISO 00f), and will be supplied with accompanying information corresponding to EN 1041 (CEN 98) or a future international equivalent.

7. Manufacturers will be required to report adverse events according to the rules established by the Global Harmonization Task Force (see Appendix 4) and to carry out post-market surveillance of identified devices.

Prospects and problems

The system outlined above is very similar to the EC Medical Devices Directive and might be described as a logical development of it. As described in the body of this book, the regulatory systems now in force, or planned for early introduction, in Australia, Canada, EC, Japan, USA and several countries in South-East Asia and Eastern Europe incorporate some, if not all, of its features. It would be straightforward for a system such as this to be adopted in most of the major countries of the world and we may expect to see a steady progress of development of national regulations towards the form described in the previous section.

Nevertheless, several difficulties would need to be overcome for such a system to be implemented to a point at which devices bearing the mark of conformity would be accepted everywhere as satisfactory for marketing and use. These difficulties will be discussed:

- In many administrations legal changes would be needed to allow marketing and use of products which have not been specifically examined and

---

[1] 'Verification' is defined in ISO 8402 (ISO 94d) as: confirmation by examination and provision of objective evidence that specified requirements have been fulfilled.
[2] 'Validation' is defined in ISO 8402 as: confirmation by examination and provision of objective evidence that the particular requirements for a specific intended use are fulfilled.

approved by the national regulatory authority. This is apparent from the Mutual Recognition Agreement between the EC and the USA (EC 98i, USA 98c). This Agreement, which covers a wide range of products, states in Articles 3.1 and 3.2 that each party 'shall, as specified in the Sectoral Annexes, accept or recognize results of specified procedures, used in assessing conformity to specified legislative, regulatory and administrative provisions . . . produced by the other Party's conformity assessment bodies and/or authorities' but the Sectoral Annex on Medical Devices (Articles 11 and 12) reduces this to the provision of reports which 'will normally be endorsed by the importing Party'. Apart from the legal issue, this situation reflects the lack of confidence in third parties which is felt in many administrations—even in some Member States of the EC as discussed in Chapter 3. It is likely that the marking alone will not be accepted as evidence of conformity, at least not until after many years of satisfactory experience, and reports or dossiers will also be needed although these may be given only cursory examination (see also Dickinson (1997)). Merrill (1998) discusses the legal implications for the United States in more detail.

- There is unlikely to be significant disagreement with the 'essential principles' but there may be a failure to agree on a body of 'recognized standards'. The mechanism for addressing the recognition of standards already exists to some extent. Working Group 2 of ISO Technical Committee 210 has considered this issue and has drafted ISO/CD 16142 (ISO 99b) which includes an Annex listing international basic safety standards and group safety standards (see Chapter 7) corresponding to individual essential principles. A clearly defined link with the Global Harmonization Task Force is needed to involve the regulatory authorities in this process and to extend it to the recognition of product safety standards. It has already been proposed that the Annex to ISO/CD 16142 should be frequently updated and should be made available on the world wide web. This process should result in a body of standards which will be generally accepted although most authorities will have a subsidiary list of national standards which will decline as more international standards of acceptable quality are produced by according to the guidance given in ISO/IEC Guide 63 (ISO/IEC 99a).

- There has been little or no work done to establish international agreement on the clinical evaluation of medical devices. It is generally believed that the differences between the major administrations—particularly between the EC and the USA—in this respect are too great for any agreement to be reached. This impression has been fostered by several publications in the USA (e.g. Nordenberg (1996), Hastings (1996), Kennedy (1996), USA 95b, USA 96a) which have misrepresented the EC requirements by ignoring the essential requirements dealing with risks and side effects. The review in Chapter 8 shows that in terms of the determination of the devices for which clinical evaluation is needed to show compliance with the essential principles, the way in which clinical investigations are to be conducted, and the assessment of outcomes for risk/benefit analysis, there is a large measure of agreement between the EC and the USA (and other administrations). The USA is unique in requiring comparative studies of competing technologies and such studies are likely to remain as additional to any global agreement on clinical evaluation. This requirement appears to stem from a perceived difference in mission between the FDA and other regulatory authorities.

The mission of a regulatory authority

The task of any regulatory authority is to enforce the laws and regulations governing medical devices. As discussed in the body of this book, such laws and regulations generally require the devices to be safe and effective or to be safe and to perform as indicated by the manufacturer. It is in the interpretation of the requirements that the mission of the regulatory authority emerges.

In the UK Department of Health and Social Security (as it was some twenty years ago), the mission was understood to be 'to ensure that medical devices are safe and fit for their purpose'. 'Their purpose' was seen as that claimed by the manufacturer. 'Safety' embraced the patient, the user (if not the patient), other persons and the environment. The regulatory activity was confined to ensuring that devices were safe and worked as claimed. Issues relating to the choice of device were regarded as being for the user or purchaser of the device. Indeed, the DHSS wished to make available to the National Health Service the greatest possible variety of safe medical devices and, when new requirements such as the Manufacturer's Registration Scheme (see Chapter 2) were introduced, the DHSS took on a supportive role in helping manufacturers to meet the requirements in order to maintain this variety.

The EC Medical Devices Directive was intended to embody a similar mission and the sixth paragraph of the preamble to the Directive states: 'Whereas medical devices should provide patients, users and third parties with a high level of protection and attain the performance levels attributed to them by the manufacturer...'. As long as such protection is achieved and the device performs as claimed, the Directive and its enforcement agencies play no part in the choice of devices or in providing information relating to the choice of devices.

The US Food and Drug Administration has adopted a different mission. As reported in Chapter 8, the Temple Report (USA 93c) took the view that clinical investigations of new medical devices should make it possible to determine whether a new device was as effective as already available therapies. Dr Bruce Burlington, then Director of the Center for Devices and Radiological Health, is quoted (Nordenberg 1996) as saying 'we must bear in mind that our goal is not simply to get products to market, but to get products that work and that we know how to use.' Under these influences, the FDA has adopted a mission of giving doctors information on the comparative advantages of alternative technologies and advice on how to use new technologies. While this is of undoubted importance, it places a major burden on the FDA—a burden that is not encountered by other regulatory agencies and which would be better placed on an agency established for this specific purpose. In the UK, such an agency, the National Institute for Clinical Excellence, has recently been established (UK 97) with a remit to examine both the effectiveness and cost of new technologies and is expected to have a major impact on purchases of both drugs and devices by the National Health Service. If such a step were taken everywhere, the separation of these issues from those addressed by the regulatory authority would allow the regulators to concentrate on a mission dedicated to allowing to market devices which were safe and fit for their purpose, thus enabling the regulatory process to be speeded up and possibly improved.

Mutual recognition

Merill (1998) identifies five models of mutual recognition agreement (MRA) and places these in the following order:

(1) The 'agent-in-place' model under which a partner country offers to the US the results of inspections by its own inspectors; the FDA then decides, on the basis of this information, whether the device meets US requirements.
(2) An 'enforcement discretion agreement', i.e. the US agrees that it will monitor less closely products coming from a country whose domestic regulatory requirements are considered to be acceptable.
(3) The 'deputy sheriff' model under which officials of the partner country verify compliance with US requirements.
(4) The 'equivalence' model. This means that the US agrees that it accepts a partner country's requirements as being equivalent to (though not the same as) US requirements.
(5) The harmonization model according to which the requirements of the US and the partner country are the same.

The MRA between the EC and the US falls somewhere between Merill's 'agent-in-place' and 'deputy sheriff' models, but the aim of most current work in the Global Harmonization Task Force, the international standards bodies, and the World Trade Organization is to move to harmonization.

The adoption of a common regulatory system world-wide would offer appreciable benefits to manufacturers who would need to make only one model of their device and prepare only one kind of submission documentation and also to regulators who would be able to share their expertise and improve the regulatory process.

The greatest economic benefits would come, however, if a device approved in one administration were to be accepted in all the others. This demands not only a global regulatory system but confidence from each administration in the conformity assessment bodies used by the others.

It has been difficult to gain such confidence in Europe (Kent 1996) and much Commission activity has been devoted to measures aimed at improving the operation of the Notified Bodies system (see Chapter 3 and Chapter 6). At the international level, too, considerable effort is being given to addressing the issue of the competence of certification bodies of all kinds. A document on the accreditation of inspection bodies has been published recently (ISO/IEC 98) and it is known that ISO/CASCO[3] is working on several documents dealing with related issues, while the recently-formed International Accreditation Forum has yet to make its mark.

These activities may be expected to bear fruit in due course but the author's view is that, while some bilateral mutual recognition activities, such as those between the EC and several other countries (EC 98i, j, k and l) may be successful, widespread mutual recognition is likely to follow many years after the adoption of a global medical device regulatory system.

---

[3] CASCO is the ISO Committee on Conformity Assessment.

Chapter 11

Overview and look to the future

The current situation

This book has reviewed the development of regulations intended to ensure that public health and safety is maintained when medical devices are used. It is only 25 years since the world's first comprehensive and rational regulatory system for medical devices (the US Medical Device Amendments of 1976 (USA 76a)) was introduced and became the model for much national legislation.

The publication of these Amendments was a landmark act which has left its mark on all subsequent medical device regulations but it must be remembered that these Amendments were made to the Food, Drug and Cosmetic Act (USA 68) and show evidence of their derivation from drug regulations which are, in some instances, inappropriate for medical devices (Samuel 1994). Much of this book has been devoted to the thesis that medical devices are engineered products which are better regulated if treated as such.

Today's model is the EC Medical Devices Directive of 1993 (EC 93a) which broke away from the pharmaceutical approach and introduced several new features appropriate to engineered products. This Directive is described in some detail in Chapter 3, and is compared with other national regulations in Chapters 4 and 5. The introduction of this Directive, which makes use of quality systems for conformity assessment, at about the same time as the Japanese and US administrations were moving towards the adoption of design controls, led to a new era in international cooperation and the publication of several documents of great significance by the Global Harmonization Task Force.

There is a movement world-wide for countries (sometimes combined into regional groupings) to introduce new or revised medical device regulations. These are generally modelled on some existing system which, until recently, was the US MDA but which is increasingly the EC MDD. The Global Harmonization Task Force is attempting to take the best features of these (and other) systems and to issue documents which will eventually constitute the building blocks of a universal regulatory system which is likely to resemble that described in Chapter 10.

The significant steps

The use of standards

Following the US Medical Device Amendments of 1976, the most significant event influencing the development of medical device regulation was the move in the European Community to complete the single European market by 1992. As is described in Chapter 2, this movement occurred at a time when advances in medical device technology had left behind the rudimentary regulations existing in European countries and medical device manufacturers had formed themselves into associations powerful enough to press for uniformity within Europe and to have a voice in the form of the new regulations. The part taken by the industry in the development of the medical devices Directives was crucial in the move from the drug regulation model. This was because the medical device industry was, with few exceptions, quite separate from the drug industry and because manufacturers had struggled for years with the attempts of regulators to apply inappropriate drug laws to their products. US manufacturers, for whom Europe was an important market, were keen to offer their experience of the Medical Device Amendments and of what they saw as its drawbacks. The European Commission lacked both numbers and expertise sufficient to develop the required Directives and seized the help offered by the industry which was accustomed to designing and manufacturing medical devices by methods quite different from those used by the drug industry.

The timing of the developments in Europe was such that other factors were influential. By the mid-1980s, several IEC 601 standards had appeared and had made their mark. Several important standards had been issued by ISO, and CEN and CENELEC, which generally adopted international standards for use in Europe, were trying to find their place in the schemes of the European Commission. At that time, the European countries were supplying most of the expertise—both in industry and public authorities—found in the international standards bodies and several were using national standards derived from ISO and IEC standards in their regulatory approaches.

The scene was therefore set for a greater use of standards in the New Approach Directives, and countries such as Germany and UK, which recognized that at least some types of medical device were not pharmaceuticals and which had separate government departments for the control of drugs and (some) devices, were able to send device experts to the Commission working groups. These experts not only pressed for the adoption of the New

Approach (which was designed for industrial products) for the medical device Directives, but resisted attempts to add an 'effectiveness' requirement into the regulations in the belief that 'performance' was a characteristic of the device alone which could be determined by objective tests. This ability to characterize and test devices by objective methods is a further feature distinguishing devices from drugs.

The use of quality systems

The publication of the Council Decision on the New Approach (EC 85b) was followed by the publication of the first edition of ISO 9001 in 1987. This standard, giving an agreed international form to the growing recognition of quality systems in the design and manufacture of goods of all kinds, rapidly assumed enormous importance in the European regulatory system. The Council Resolution on the global approach (EC 89a) and the Council Decision on the modules for conformity assessment (EC 90a) (see Chapter 2) laid great emphasis on the use of the ISO 9000 series in New Approach Directives. This emphasis was welcomed by the medical device industry, large parts of which were already subject to the US Good Manufacturing Practice Regulation (USA 78) and the UK Manufacturers' Registration Scheme. The critical step taken by the European legislators was to use compliance with a quality system standard as part of the conformity assessment process. This contrasted with the situation in the United States where compliance with GMP was a post-market check, and gave a regulatory framework to the approach pioneered in the UK with its (advisory) Manufacturers' Registration Scheme.

The use of quality systems as a key feature of conformity assessment received a boost when the second edition of ISO 9001 (ISO 94a), with its greatly expanded clause on design control, was published in 1994. As described in Chapter 6, this led to the revision of the Japanese and US GMP Regulations to bring them into alignment with the international standard, and began movements in both these countries to make compliance with the quality system regulations a requirement of the pre-market approval process. These steps gave tacit recognition to the fact that medical devices are designed and manufactured like other engineering products.

The appearance of 'validation'

Chapter 8 discusses at some length the arguments that have raged, and are still carried on, over the question of effectiveness requirements for medical devices and the place of clinical investigations. As mentioned above, the experts involved in the development of the EC medical device Directives broke away from the drug pattern in making the requirement one related to 'performance', coupled with a risk/benefit assessment where necessary.

As discussed in that chapter, the differences between the approaches of the EC, Japan and the US, and the results of their processes, are, when they are examined in detail, much less marked than has often been alleged—the exception being the comparative studies of outcomes carried out in the US.

There can be no argument over the fact that medical devices must not only be safe but must act in a known, precise and reproducible manner that meets some need of the health professional and/or patient. Showing that the needs of the health professional/patient are met is the process identified in the ISO 9001 quality system standard as 'validation' and the author's view is that the time for dispute about whether effectiveness or performance is the right property to be assessed is past and that the discussion should be about the most appropriate ways of carrying out validation. This will in most cases involve a clinical evaluation of the safety and behaviour of the device under its normal working conditions but will demand a specific clinical investigation only where some novelty is present—the European Notified Bodies Group has put forward some criteria for clinical investigations which should be taken forward and developed by the GHTF.

The kinds of investigation that compare different devices and even procedures in order to give the doctor guidance on which device/procedure to use in a given situation, or to give the reimbursement agency advice on which device/procedure is most economic, are outside the remit of most regulatory agencies. Most developed countries have established bodies independent of the regulatory authority for these purposes.

The recognition of the limits of PMA

In spite of the efforts made to design regulatory systems that will ensure that all medical devices are safe, problems still occur. The numbers of adverse effects experienced are high—more than 6000 in the UK in 1998 (UK 99)—but very few are serious.

For many years most developed countries have had mechanisms of some kind for the reporting of adverse incidents with medical devices but the Medical Device Reporting Regulations (USA 84), published in the US in 1984, was the first introduction of a comprehensive legal reporting requirement. The enormous numbers of reports submitted to the FDA under these regulations helped to bring about a realization that the pre-market approval processes could never eliminate problems. Similar requirements were inserted in the EC medical devices Directives and have been incorporated in almost all recent medical device legislation.

The widely-publicized problems with pacemakers, heart valves and breast implants, as well as the large numbers of less well-known incidents, have led to increased attention being given to corrective action (note, for instance, the strengthened requirements in Clause 4.14.2 of ISO 13485—particular requirements for the application of ISO 9001 to medical devices (ISO 96a)), to the need for continual analysis of reports of all kinds in order to detect systematic failures at the earliest opportunity (post-market surveillance requirements are found in the EC, Japanese and US regulations—see Chapter 9), and to maintaining information about the location of sensitive devices such as implants (the tracking requirements in the US, as modified by the FDA Modernization Act of 1997 (USA 97a) are the most demanding example of such information gathering).

The Global Harmonization Task Force

Possibly the most significant of the events identified in this section has been the formation of the Global Harmonization Task Force (GHTF) in 1992. Up to then, there had been little contact between regulatory agencies and attempts at formal cooperation were limited to the Canada/UK/US Tri-Partite Working Group which had produced a joint scheme for bio-compatibility assessment (which later became the basis of ISO 10993 (ISO 94e)) and a Memorandum of Understanding between the UK Department of Health and Social Security and the US Food and Drug Administration on GMP compliance. Japan, in particular, was regarded as being remote and insular.

Nevertheless, Japan joined Canada, EC and US in forming the GHTF and has been not only an active participant but a user of its outputs. Although the initial objectives of the GHTF, which has been expanded to include Australia, were limited to harmonizing quality systems, it moved rapidly to address many other aspects of medical device regulation with a remarkable degree of success.

As Hilz (1997) puts it, 'Would you have ever anticipated five or six years ago that members of regulatory agencies and representatives from industry in the three main trading blocks would sit together in a relaxed and open-minded atmosphere and discuss in a straightforward way national regulations with the aim of harmonizing them?... Would you have imagined that this group of people would come to results in a surprisingly short period of time?'

The amazing success of the GHTF probably results from the small size and informality of the Study Groups. The regulators have been surprised to find that regulatory approaches different from their own have rational bases and have learned to abandon defensive attitudes towards their own systems. Dialogue between industry and authority representatives has taken place without the usual confrontation, and mutual respect has grown.

The list of documents emanating from the GHTF, given in Chapter 10, is impressive and is being added to steadily. The degree of acceptance of these documents, not only by the six members but by many other countries in the process of introducing or amending medical device regulations, has been higher than could have been expected. Cooperation with ISO Technical

Committee 210 remains important and can be expected to increase following the signing of a Memorandum of Understanding between the two bodies during 1999.

Recent developments

Agreement on essential principles, the use of standards, the appropriate quality systems, the classification of devices and adverse event reporting has almost been reached among the leading administrations, responsible for the production and use of the majority of medical devices. Adoption of the format for the documentation of approval submissions, which is now being worked on by the Global Harmonization Task Force (GHTF 99), is likely soon and this will greatly reduce the time spent in preparing submissions for medical device approvals. These key GHTF documents have been issued for comment by the FDA (USA 98d, USA 99a) and acceptance in the USA would greatly support the global harmonization movement.

The US appears to have completely changed its attitude to voluntary standards. It may be recalled that the Medical Device Amendments of 1976 (USA 76a) identified Class II devices as those for which compliance with performance standards was necessary, in addition to the general controls applied to Class I devices. This provision was never applied as the FDA would only consider standards developed in-house. During the 1970s and 1980s the FDA had a deep distrust of voluntary standards bodies, including ISO and IEC, which it regarded as 'manufacturers' clubs'. It refused to accept that their standards would provide adequate patient protection and insisted that it was the FDA's responsibility to produce standards for use in the regulatory process. The effort needed to develop proper standards proved to be too great and the standards provision for Class II was never applied.

Gradually this attitude changed as FDA staff began perhaps first to appreciate the value of the IEC 601 series of standards and then to participate in ISO and IEC Technical Committees. Claims of compliance with standards from recognized bodies (both American and international) became features of pre-market approval submissions and the FDA Modernization Act of 1997 (US 97a) provided for the 'recognition' by the FDA of voluntary standards and for the acceptance of manufacturers' declarations of compliance. There are now more than 500 recognized standards (164 international) listed on the FDA website.

This new attitude to standards is linked with a recognition, emanating from work in the Global Harmonization Task Force, that scientists, whatever their background or situation, in attempting to decide on the safety and acceptability of new devices will seek the same features, will work from a formal or informal check list of 'essential requirements' and will generally arrive at the same conclusions. This has been demonstrated by the ready acceptance of the 'essential principles' issued by the GHTF (which are almost identical with the essential requirements listed in Annex I of the Medical Devices Directive (EC 93a)) and the publication by ISO TC210 of a list of standards linked to those essential principles (ISO 99b).

The growing importance of risk management has been recognized by the recent publication of ISO 14971 (ISO 00b), the proposal to incorporate a reference to this standard in the forthcoming new edition of IEC 601-1, the moves within CEN to replace EN 1441 (CEN 97c) by the international standard, and the commencement of work within GHTF Study Group 3 to relate this standard to the regulatory process.

Not all the recent developments are encouraging, although they may ultimately lead to positive developments. The introduction by France of its law (France 98) requiring the submission of data to the Competent Authority three months before placing certain devices on the market is, on the face of it, a threat to the entire system of the EC medical devices Directives but could lead to a recognition that the Directives need to be strengthened in certain respects. If such a strengthening—which relates to the consistent and thorough operation of the conformity assessment procedures by the Notified Bodies—could be achieved, it would lead not only to an increase of confidence within the EC, but to a smoother and more rapid operation of the mutual recognition agreements between the EC and other countries. The review, and expected revision, of the MDD, now under way, should address these and other shortcomings.

At present, the MRA between the EC and the US is suffering from a lack of confidence by the FDA in the European Conformity Assessment Bodies (CABs). A letter from the FDA (USA 99b) points out the need for training of European CABs in the requirements of the FDA which, they are at pains to point out, go beyond quality system assessment to include checking compliance with other US regulations such as Medical Device Reporting, Tracking, etc. These are matters which should be capable of resolution, perhaps by further development of the GHTF auditing guidelines, but illustrate a more fundamental distrust by the FDA of private sector third party assessment. Successful operation of the Notified Bodies Oversight Group should overcome this lack of confidence.

Although a pilot programme of third-party assessment was announced in 1996 (USA 96c) and extended by provisions of the FDA Modernization Act of 1997 (USA 97a), this programme has been limited to a small number of devices and a small number of third parties. As the programme involves a financial penalty for manufacturers opting for a third party review (as fees must be paid, whereas FDA review is free of charge) as well as suspicion that the FDA will not trust the results, there has been poor take-up of a programme which is needed if confidence in third party review is ever to be gained.

Another issue which demands attention is the recent publication of the revised version of the ISO 9000 series of quality standards. ISO 9001:2000 (ISO 00d) will replace the current ISO 9001, 9002 and 9003, during a three-year transition period, allowing flexibility in the application of the standard to cope with the different levels of the quality system. More fundamental is likely to be the redirection of the document to address customer satisfaction and continuous improvement. Although ISO TC 210 is already working on revisions to ISO 13485/13488 (ISO 96a, ISO 96b) to adapt them to the new parent standard, these are not expected to be published before 2002 and medical device manufacturers and the regulatory authorities will need all this time to adapt their procedures to the new standard.

The next decade

For many years medical device manufacturers operating in a world-wide market place have dreamed of a universal regulatory system which would allow them to market everywhere a device which had been through the regulatory process in one administration. The benefits offered by such a system are described by George (1998) as:

- improved patient access to new products following from the elimination of delays in obtaining multiple approvals and of the use of regulations for protectionist purposes
- reduction in time and cost spent by manufacturers on multiple submissions, and hence a reduction in the cost of medical devices
- improved availability of better and cheaper devices resulting from world-wide competition.

To these, the author would add:

- more rapid detection and correction of problem devices as a consequence of global adverse effect reporting and post-market surveillance.

This somewhat utopian situation may never be achieved but progress towards it is already being made and several of the major elements of a global regulatory system will be in place probably during the next five years. The regulatory authorities themselves recognize the benefits that could accrue to their operations in terms of more efficient use of their expertise and the opportunity to concentrate on the issues arising from new technological developments if much of the routine work could be made redundant.

The outstanding areas in which agreement is lacking are the conformity assessment processes appropriate to each class of device (the author's view is that this is not likely to be too difficult); the evaluation of clinical data; the need for clinical investigation, and the conduct of risk/benefit assessments.

Even acknowledging that these areas of difficulty remain and that progress in resolving them may be limited—and it must be borne in mind that these have not yet been attempted—and even if some countries remain outside the influence of the global harmonization process, it has to be recognized that the convergence of views that has occurred in the last five years is remarkable. The emergence of the trends identified earlier is clear. The pace of activity in the Global Harmonization Task Force and the international standards bodies is such that further movement towards the widespread adoption of elements of the 'ideal' system I have described is assured.

We are at a point where the difference between 'harmonization' and 'mutual recognition' is becoming marked. Some years ago many workers in the field of medical device regulation would have said that unifying the world's regulatory systems was an impossible task but that, as no one system could be seen as demonstrably better than the others, mutual recognition of approvals should be achievable. In fact, the reverse now seems to be the case. As pointed out above, national and regional regulatory systems are becoming closely aligned and are being based on the same international standards for products and quality systems. Within a few years we can expect to see an approximate global harmonization.

Such a system would work and would offer benefits irrespective of whether the conformity assessment processes were carried out in each country/region by third parties or by the regulatory authorities themselves. However, approvals given under such conditions would apply only in the country/region in which they were obtained. The desired acceptance worldwide of approvals given anywhere is dependent on confidence in conformity assessment carried out elsewhere and is unlikely to be achieved during the next decade, although by that time we should be on the point of mutual recognition of approvals between several major players.

Several further developments are likely to assist both the harmonization process and the gaining of mutual confidence:

–   The European Community is likely to make some moves to strengthen the system for assessment of Class IIb implants and Class III devices in response to the French Law on the surveillance and control of certain products (France 98). These moves may include tighter supervision of the designation and operation of Notified Bodies and allowing Competent Authorities a larger part in the approval process. The Mutual Recognition Agreement between the EC and the USA (EC 98i) will be a further influence on the EC to improve the operation of the Notified Bodies. The author would be pleased to see the application of quality systems become a mandatory feature of the EC conformity assessment process, rather than an option as it is at present.
–   In the United States we are likely to see an increasing use of standards, possibly leading to a reaffirmation of the Class II process (compliance

with standards) originally incorporated in the 1976 Medical Device Amendments (USA 76a) and the abandoning of the 'substantially equivalent' provisions of Section 510(k). In discussion at the 7th Global Medical Device Conference, Sydney, February 1998, FDA staff indicated that they had been prepared to make this step part of the FDA Modernization Act of 1997 (USA 97a) but had been prevented by other considerations.

– In Japan further steps towards the acceptance of recommendations from the Global Harmonization Task Force can be expected.

– In the rest of the world, as well as in the 'Big Three', the advantages of adopting the GHTF documents will become increasingly apparent. Manufacturers will press for the use of the GHTF technical file which will be developed and improved in the light of experience of its use. New 'key factors' may emerge (risk analysis/management being one) and new, wide-ranging standards from ISO/TC 210 will assist their emergence. Work by the Notified Bodies Group in Europe on the evaluation of clinical data is likely to move into the GHTF and/or ISO/TC 210 and to lead to a better understanding of this difficult process. We can expect the gradual emergence of a general approach to the evaluation of clinical data for safety, performance and risk/benefit assessment, with the USA (and, possibly, other countries) requiring additional studies relating to the choice of competing technologies or procedures.

– The process of regionalization in the Far East and South America should be encouraged by allowing membership of the Global Harmonization Task Force to the Asia–Pacific Harmonization Group and MERCOSUR (see Chapter 5).

The gaining of mutual confidence in the conformity assessment bodies will be difficult and slow, but the convergence of the regulatory systems will make comparisons easier and will lead to a faster adoption of best practice. The emergence of a more efficient, more effective, less varied, world-wide regulatory system for medical devices is within sight only 25 years since the first real system (USA 76a) was introduced.

Appendix 1

Practical steps towards global harmonization

Introduction

The Council resolution of 21 December 1989 on a global approach to conformity assessment refers to the promotion of international trade by concluding mutual recognition agreements with the Community's trading partners. The object of the agreements is the mutual recognition of certificates, marks of conformity and test reports issued by either Party. This paper deals with some issues relating to the practical implementation of this philosophy in the case of medical devices.

The approaches to medical device regulation differ considerably in the three major market places (Europe, USA, Japan), but a requirement for quality assurance in the design and/or manufacturing processes is a feature of all. While the mutual recognition of medical device approvals may, therefore, be a distant goal, the reciprocal acceptance of quality assurance systems as an intermediate step towards this goal is a realistic objective that would offer significant benefits to both manufacturers and authorities.

The achievement of this objective will demand determination and a recognition by all parties that some compromise and departures from established positions may be necessary. The Commission has built considerable flexibility into its proposal for a medical device directive; the achievement of reciprocal acceptance of quality systems and, eventually, mutual recognition of product approvals will be difficult, if not impossible, if other parties are locked into regulatory systems which require rigid adherence to detailed procedures.

1. Building on ISO 9000

The acceptance of the ISO 9000 series of quality system standards in the USA, Europe and Japan is the key to the reciprocal acceptance of quality systems.

The need for supplementary requirements to relate the ISO standards specifically to medical devices is accepted worldwide. These supplementary requirements must be similar or identical in their main features if our goals are to be achieved.

It is, therefore, essential that those responsible for developing the EN 46000 series of supplementary requirements, the US GMP Regulation and Japanese quality system requirements collaborate to eliminate differences. A detailed examination of the various drafts is necessary and we urge the exchange of documents between participants at the earliest possible stage. The preparation of the EN 46000 series in Europe is a relatively open process and comments from the USA have been received and considered. However, in the USA the quality system requirements will take the form of a Regulation and it does not appear to be possible for a European input to be made in a similar way. Furthermore, the process by which these requirements will be implemented in Japan, and how a European input can be made, is not known.

Possibly as important as the standards and regulations themselves, will be the body of supporting documents that will be developed to elaborate the requirements and assist in their interpretation. The following guidance documents are currently in preparation in Europe:

prEN 50103   Guidelines on the application of ISO 9001/EN 46001 and ISO 9002/EN 46002 for the active (including active implantable) medical device industry.

prEN 724   Guide to the application of EN 29001/EN 46001 and EN 29002/EN 46002 to the manufacture of non-active medical devices.

CEN TC140 N90   Guidance on the application of EN 29001/EN 46001 and EN 29002/EN 46002 for the in-vitro diagnostics industry.

These documents are available for review and comments on them can be submitted to CEN/CENELEC via the national standards organizations in the USA and Japan.

It is well known that the FDA have compiled many guidance documents for use by their inspectors. To assist the harmonization process these guidance documents should be made available to the other parties and their preparation should become a more open process. Alignment of guidance documents or, better still, collaboration in the preparation of one set of guidance, is seen as crucial to the achievement of mutual recognition of quality systems.

2. Practical issues in the harmonization of quality system requirements

Alignment of the basic standards and guidance is an essential foundation for mutual recognition, but several other aspects of the process of assessment must be aligned if approval of a manufacturer's quality system by one party is to be accepted by the other parties.

These include:

- documentation
- environmental (clean room) requirements
- sterilization validation
- inspection intervals and periods of validity of approvals
- responses to identified non-compliances
- appeals procedures.

The alignment of these facets of the process would clearly be facilitated by such steps as:

- universal guidance documents
- similar requirements for inspection bodies
- common training programmes for inspectors.

3. Problem areas

Several areas of difficulty have already been identified. However, the ways of addressing these are clear and they can be overcome by application and goodwill. Two other areas of difficulty appear to be more serious:

(a) In the USA, the requirements for quality systems are decided by a Government agency and the inspections are carried out by the same agency. In Europe, there is no central body corresponding to the FDA and the setting of the requirements is delegated to the standards bodies and the inspections are delegated to Notified Bodies.

There is clearly a challenge for the USA to accept this degree of delegation to non-governmental bodies. This may be easier if the extent of the due process involved in creating a European standard is fully understood and if the role of the Competent Authorities in accrediting and supervising the Notified Bodies is recognized.

(b) The Commission proposal for a directive on medical devices envisages the use of the three ISO standards 9001, 9002, 9003 according to the classification of the medical device. It is our understanding that the revised US GMP regulation is likely to require a quality system equivalent to ISO 9001 for all medical devices. Such a divergence of approach will certainly create difficulties and it is recommended that the US authorities recognize the existence of the three ISO standards and introduce some flexibility into their Regulation.

4. Fundamental obstacles to full mutual recognition

This paper presents a technical discussion of quality systems and the prospects for reciprocal acceptance. This appears to be possible and worthwhile but it must be emphasized that quality systems form only part of the approval process and that fundamental obstacles to the ultimate goal of mutual recognition of product approvals will remain even if such reciprocal acceptance is achieved.

When the medical device directives come into effect, the fundamental legal requirement in Europe will be compliance with essential requirements for safety, performance and labelling. These essential requirements will be elaborated in harmonized voluntary standards. A quality system to ISO 9001 will be an important element of the proof of compliance with the essential requirements (for some products) but other methods will also be permitted.

In the USA, the fundamental requirement is for substantial equivalence with an existing device known to be satisfactory or for proof of safety and efficacy to the satisfaction of the FDA. The level of proof is based on non-explicit criteria set by the FDA and this process can present great obstacles to non-US manufacturers. It is believed that compliance with quality system requirements equivalent to ISO 9001 will become an essential pre-requisite of approval.

In Japan, compliance with the Pharmaceutical Affairs Law requires that a foreign manufacturer obtains both a product approval and an import licence from the Ministry of Health and Welfare. The criteria on which decisions are based, and the extent to which non-Japanese data are accepted, are unclear. Compliance with GMP requirements is part of the process.

More transparency and objectivity in the US and Japanese approval processes, as well as a willingness to allow established systems to evolve in line with international developments, will be necessary if full mutual recognition is ever to be achieved.

5. Next steps

The Commission welcomes the opportunity presented by the Nice meeting to discuss these issues with colleagues from other continents. We see the need for an informal group to continue discussion in order that the alignment of approaches to quality assurance and the overcoming of some of the problems identified in this paper can be achieved.

In particular, we see an immediate need to begin work on two tasks:

- the alignment of supplementary requirements to ISO 9001/2
- the alignment of guidance documents.

We believe these should be addressed by a Task Force composed of representatives from the three parties: Europe, USA and Japan. In the longer term, the handing over of these activities to the international standards bodies must be considered. This hand-over would demand a commitment by all parties to accept and adopt the resulting international standards.

These first tasks need to be followed by joint activities in the practical issues identified in Section 2. Some of these are already the subject of international standards activities. Full participation in these activities by all parties and commitment to the acceptance and use of the standards is essential.

Progress on these tasks and the general movement towards global harmonization should be monitored by a Forum consisting of representatives of the industries and administrations of the three parties. The Commission is ready to commit to the support of, and participation in, such a Forum which might meet twice-yearly and seeks similar commitments from all the others involved.

The commencement of harmonization activities in the field of quality assurance is foreseen as leading to similar activities in other areas such as:

- the provision of clinical data as part of the approval process
- 'vigilance', or incident reporting, requirements
- post-market surveillance, or tracking, requirements.

The establishment of a Forum on the lines described, if it receives the full support of the administrations, offers the prospect of a continuing activity dedicated to the removal of obstacles to mutual recognition and the promotion of international trade and patient benefit.

Appendix 2

Essential principles of safety and performance of medical devices

Endorsed by The Global Harmonization Task Force

FINAL DOCUMENT

30 June 1999

The document herein was produced by the Global Harmonization Task Force, a voluntary group of representatives from medical device regulatory agencies and the regulated industry. The document is intended to provide non-binding guidance to regulatory authorities for use in the regulation of medical devices, and has been subject to consultation throughout its development.

The Global Harmonization Task Force: GHTF.SG1.N020R5

Foreword

Study Group 1 recognizes that to further the processes of global harmonization of regulatory requirements, it is necessary to have common guidelines to indicate the Essential Principles of safety and performance of medical devices in the interests of public health.

Existing regulations and draft regulations of participating members of the Task Force have already included, in many cases, such statements of principles and it has been the conclusion of the Study Group that, although presented in different ways, the features of such principles are common to all such regulations.

For these reasons, Study Group 1 proposes that the following set of principles should be considered in the development or amendment of regulatory systems.

The Study Group recommends that Regulatory Authorities consider accepting, for placing on the market, medical devices where compliance with relevant Essential Principles has been demonstrated.

Notes

(1) There may be further safety and performance principles for devices incorporating substances derived from tissues of human or animal origin and in vitro diagnostic devices. This may suggest the need for additional review of this Essential Principles document in the future.

(2) It is understood that the operation of a quality system, the use of standards, post-market vigilance, the pre-market review of a technical file, type testing and final product testing, are all important means, which may individually or jointly be utilized, to achieve compliance with the Essential Principles. These matters are not addressed within this document.

(3) Information on labelling requirements is the subject of a separate document.

GLOBAL HARMONIZATION TASK FORCE
STUDY GROUP 1

ESSENTIAL PRINCIPLES OF SAFETY AND PERFORMANCE OF MEDICAL DEVICES

General requirements

1.   Medical devices should be designed and manufactured in such a way that, when used under the conditions and for the purposes

intended and, where applicable, by virtue of the technical knowledge, experience, education or training of intended users, they will not compromise the clinical condition or the safety of patients, or the safety and health of users or, where applicable, other persons, provided that any risks which may be associated with their use constitute acceptable risks when weighed against the benefits to the patient and are compatible with a high level of protection of health and safety.

2. The solutions adopted by the manufacturer for the design and construction of the devices should conform to safety principles, taking account of the generally acknowledged state of the art. In selecting the most appropriate solutions, the manufacturer should apply the following principles in the following order:

- identify hazards and the associated risks arising from the intended use and foreseeable misuse,
- eliminate or reduce risks as far as possible (inherently safe design and construction),
- where appropriate take adequate protection measures including alarms if necessary, in relation to risks that cannot be eliminated,
- inform users of the residual risks due to any shortcomings of the protection measures adopted.

3. Devices should achieve the performance intended by the manufacturer and be designed, manufactured and packaged in such a way that they are suitable for one or more of the functions within the scope of the definition of a medical device applicable in each jurisdiction.

4. The characteristics and performances referred to in Clauses 1, 2 and 3 should not be adversely affected to such a degree that the clinical conditions and safety of the patients and, where applicable, of other persons are compromised during the lifetime of the device, as indicated by the manufacturer, when the device is subjected to the stresses which can occur during normal conditions of use and has been properly maintained in accordance with the manufacturer's instructions.

5. The devices should be designed, manufactured and packed in such a way that their characteristics and performances during their intended use will not be adversely affected during transport and storage taking account of the instructions and information provided by the manufacturer.

6. The benefits must be determined to outweigh any undesirable side-effects for the performances intended.

Requirements regarding design and construction

Chemical, physical and biological properties

7.1   The devices should be designed and manufactured in such a way as to ensure the characteristics and performance referred to in Section I of the 'General Requirements'. Particular attention should be paid to:

- the choice of materials used, particularly as regards toxicity and, where appropriate, flammability,
- the compatibility between the materials used and biological tissues, cells and body fluids, taking account of the intended purpose of the device.
- the choice of materials used should reflect, where appropriate, matters such as hardness, wear and fatigue strength.

7.2   The devices should be designed, manufactured and packed in such a way as to minimize the risk posed by contaminants and residues to the persons involved in the transport, storage and use of the devices and to the patients, taking account of the intended purpose of the product. Particular attention should be paid to the tissues exposed and to the duration and frequency of exposure.

7.3   The devices should be designed and manufactured in such a way that they can be used safely with the materials, substances and gases with which they enter into contact during their normal use or during routine procedures; if the devices are intended to administer medicinal products they should be designed and manufactured in such a way as to be compatible with the medicinal products concerned according to the provisions and restrictions governing these products and that their performance is maintained in accordance with the intended use.

7.4   Where a device incorporates, as an integral part, a substance which, if used separately, may be considered to be a medicinal product/drug as defined in the relevant legislation that applies within that jurisdiction and which is liable to act upon the body with action ancillary to that of the device, the safety, quality and usefulness of the substance should be verified, taking account of the intended purpose of the device.

7.5   The devices should be designed and manufactured in such a way as to reduce to a minimum the risks posed by substances that may leach from the device.

7.6   Devices should be designed and manufactured in such a way as to reduce, as much as possible, risks posed by the unintentional ingress or egress of substances into or from the device taking into account the device and the nature of the environment in which it is intended to be used.

Infection and microbial contamination

8.1   The devices and manufacturing processes should be designed in such a way as to eliminate or reduce as far as possible the risk of infection to the patient, user and, where applicable, other persons. The design should allow easy handling and, where necessary, minimize contamination of the device by the patient or vice versa during use.

8.2.1   Tissues of non-human origin as far as considered a medical device, should originate from animals that have been subjected to veterinary controls and surveillance adapted to the intended use of the tissues. National regulations may require that the manufacturer and/or the Competent/Regulatory Authority should retain information on the geographical origin of the animals. Processing, preservation, testing and handling of tissues, cells and substances of animal origin should be carried out so as to provide optimal safety. In particular safety with regard to viruses and other transmissible agents should be addressed by implementation of validated methods of elimination or viral inactivation in the course of the manufacturing process.

8.2.2   In some jurisdictions products incorporating human tissues, cells and substances may be considered medical devices. In this case, selection, processing, preservation, testing and handling of tissues, cells and substances of such origin should be carried out so as to provide optimal safety. In particular safety with regard to viruses and other transmissible agents should be addressed by implementation of validated methods of elimination or viral inactivation in the course of the manufacturing process.

8.3   Devices delivered in a sterile state should be designed, manufactured and packed in a non-reusable pack and/or according to appropriate procedures to ensure that they are sterile when placed on the market and remain sterile, under the storage and transport conditions laid down, until the protective packaging is damaged or opened.

8.4   Devices delivered in a sterile state should have been manufactured and sterilized by an appropriate, validated method.

8.5   Devices intended to be sterilized should be manufactured in appropriately controlled (e.g. environmental) conditions.

8.6   Packaging systems for non-sterile devices should keep the product without deterioration at the level of cleanliness stipulated and, if the devices are to be sterilized prior to use, minimize the risk of microbial contamination; the packaging system should be suitable taking account of the method of sterilization indicated by the manufacturer.

8.7 The packaging and/or label of the device should distinguish between identical or similar products sold in both sterile and non-sterile condition.

Construction and environmental properties

9.1 If the device is intended for use in combination with other devices or equipment, the whole combination, including the connection system should be safe and should not impair the specified performance of the devices. Any restrictions on use should be indicated on the label or in the instructions for use.

9.2 Devices should be designed and manufactured in such a way as to remove or minimize as far as is practicable:

- the risk of injury, in connection with their physical features, including the volume/pressure ratio, dimensional and where appropriate ergonomic features,
- risks connected with reasonably foreseeable environmental conditions, such as magnetic fields, external electrical influences, electrostatic discharge, pressure, temperature or variations in pressure and acceleration,
- the risks of reciprocal interference with other devices normally used in the investigations or for the treatment given,
- risks arising where maintenance or calibration are not possible (as with implants), from ageing of materials used or loss of accuracy of any measuring or control mechanism.

9.3 Devices should be designed and manufactured in such a way as to minimize the risks of fire or explosion during normal use and in single fault condition. Particular attention should be paid to devices whose intended use includes exposure to flammable substances or to substances which could cause combustion.

Devices with a measuring function

10.1 Devices with a measuring function should be designed and manufactured in such a way as to provide sufficient accuracy, precision and stability within appropriate limits of accuracy and taking account of the intended purpose of the device. The limits of accuracy should be indicated by the manufacturer.

10.2 The measurement, monitoring and display scale should be designed in line with ergonomic principles, taking account of the intended purpose of the device.

10.3 The measurements made by devices with a measuring function should be expressed in legal units as required by the legislation governing such expression of each jurisdiction in which the device is to be sold

Protection against radiation

11.1 General
11.1.1 Devices should be designed and manufactured in such a way that exposure of patients, users and other persons to radiation should be reduced as far as possible compatible with the intended purpose, whilst not restricting the application of appropriate specified levels for therapeutic and diagnostic purposes.
11.2 Intended radiation
11.2.1 Where devices are designed to emit hazardous levels of radiation necessary for a specific medical purpose the benefit of which is considered to outweigh the risks inherent in the emission, it should be possible for the user to control the emissions. Such devices should be designed and manufactured to ensure reproducibility and tolerance of relevant variable parameters.
11.2.2 Where devices are intended to emit potentially hazardous, visible and/or invisible radiation, they should be fitted, where practicable, with visual displays and/or audible warnings of such emissions.
11.3 Unintended radiation
11.3.1 Devices should be designed and manufactured in such a way that exposure of patients, users and other persons to the emission of unintended, stray or scattered radiation is reduced as far as possible.
11.4 Instructions for use
11.4.1 The operating instructions for devices emitting radiation should give detailed information as to the nature of the emitted radiation, means of protecting the patient and the user and on ways of avoiding misuse and of eliminating the risks inherent in installation.
11.5 Ionizing radiation
11.5.1 Devices intended to emit ionizing radiation should be designed and manufactured in such a way as to ensure that, where practicable, the quantity, geometry and energy distribution (or quality) of radiation emitted can be varied and controlled taking into account the intended use.
11.5.2 Devices emitting ionizing radiation intended for diagnostic radiology should be designed and manufactured in such a way as to achieve appropriate image and/or output quality for the intended medical purpose whilst minimizing radiation exposure of the patient and user.
11.5.3 Devices emitting ionizing radiation, intended for therapeutic radiology should be designed and manufactured in such a way as to

enable reliable monitoring and control of the delivered dose, the beam type and energy and where appropriate the energy distribution of the radiation beam.

Requirements for medical devices connected to or equipped with an energy source

12.1 Devices incorporating electronic programmable systems should be designed to ensure the repeatability, reliability and performance of these systems according to the intended use. In the event of a single fault condition in the system, appropriate means should be adopted to eliminate or reduce as far as possible consequent risks.

12.2 Devices where the safety of the patients depends on an internal power supply should be equipped with a means of determining the state of the power supply.

12.3 Devices where the safety of the patients depends on an external power supply should include an alarm system to signal any power failure.

12.4 Devices intended to monitor one or more clinical parameters of a patient should be equipped with appropriate alarm systems to alert the user of situations which could lead to death or severe deterioration of the patient's state of health

12.5 Devices should be designed and manufactured in such a way as to minimise the risks of creating electromagnetic fields which could impair the operation of other devices or equipment in the usual environment.

12.6 Protection against electrical risks

12.6.1 Devices should be designed and manufactured in such a way as to avoid, as far as possible, the risk of accidental electric shocks during normal use and in single fault condition, provided the devices are installed correctly.

12.7 Protection against mechanical and thermal risks

12.7.1 Devices should be designed and manufactured in such a way as to protect the patient and user against mechanical risks connected with, for example, resistance to movement, instability and moving parts.

12.7.2 Devices should be designed and manufactured in such a way as to reduce to the lowest practicable level the risks arising from vibration generated by the devices, taking account of technical progress and of the means available for limiting vibrations, particularly at source, unless the vibrations are part of the specified performance.

12.7.3 Devices should be designed and manufactured in such a way as to reduce to the lowest practicable level the risks arising from the noise emitted, taking account of technical progress and of the means available to reduce noise, particularly at source, unless the noise emitted is part of the specified performance.

12.7.4   Terminals and connectors to the electricity, gas or hydraulic and pneumatic energy supplies which the user has to handle should be designed and constructed in such a way as to minimize all possible risks.

12.7.5   Accessible parts of the devices (excluding the parts or areas intended to supply heat or reach given temperatures) and their surroundings should not attain potentially dangerous temperatures under normal use.

12.8   Protection against the risks posed to the patient by energy supplies or substances

12.8.1   Devices for supplying the patient with energy or substances should be designed and constructed in such a way that the delivered amount can be set and maintained accurately enough to guarantee the safety of the patient and of the user.

12.8.2   Devices should be fitted with the means of preventing and/or indicating any inadequacies in the delivered amount which could pose a danger. Devices should incorporate suitable means to prevent, as far as possible, the accidental release of dangerous levels of energy from an energy and/or substance source.

12.8.3   The function of the controls and indicators should be clearly specified on the devices. Where a device bears instructions required for its operation or indicates operating or adjustment parameters by means of a visual system, such information should be understandable to the user and, as appropriate, the patient.

Information supplied by the manufacturer

13.1   Each device should be accompanied by the information needed to identify the manufacturer, to use it safely and to ensure the intended performance, taking account of the training and knowledge of the potential users. This information comprises the details on the label and the data in the instructions for use, and should be easily understood.

(NOTE: Detailed information on labelling requirements is the subject of a separate document)

Clinical Evaluation

14.1   Where conformity with these Essential Principles should be based on clinical evaluation data, such data should be established in accordance with the relevant requirements applicable in each jurisdiction.

Clinical investigations on human subjects should be carried out in accordance with the Helsinki Declaration adopted by the 18th World Medical Assembly in Helsinki, Finland, in 1964, as last amended by the 41st World Medical Assembly in Hong Kong in 1989. It is mandatory that all measures relating to the protection of human subjects are carried out in the spirit of the Helsinki Declaration. This includes every step in the clinical investigation from first consideration of the need and justification of the study to publication of the results. In addition, some countries may have specific regulatory requirements for pre-study protocol review or informed consent.

(NOTE: Specific guidance on clinical evaluation may be developed in the future)

Appendix 3

Role of standards in the assessment of medical devices

Endorsed by The Global Harmonization Task Force

FINAL DOCUMENT

24 February 2000

The document herein was produced by the Global Harmonization Task Force, a voluntary group of representatives from medical device regulatory agencies and the regulated industry. The document is intended to provide *non-binding* guidance to regulatory authorities for use in the regulation of medical devices, and has been subject to consultation throughout its development.

The Global Harmonization Task Force: GHTF.SG1.N012R10

GLOBAL HARMONIZATION TASK FORCE STUDY GROUP 1

ROLE OF STANDARDS IN THE ASSESSMENT OF MEDICAL DEVICES

General principles

Principles

International standards are a building block for harmonized regulatory processes to assure the safety, quality and performance of medical devices.

To achieve this purpose, the following principles are recommended:

- Regulatory Authorities and industry should encourage and support the development of international standards for medical devices to demonstrate compliance with 'the Essential Principles of Safety and Performance of Medical Devices'[1] (referred to hereafter as the Essential Principles).
- Regulatory Authorities developing new medical device regulations should encourage the use of international standards.
- Regulatory Authorities should provide a mechanism for recognizing international standards to provide manufacturers with a method of demonstrating compliance with the Essential Principles.
- When an international standard is not applied or not applied in full, this is acceptable if an appropriate level of compliance with the Essential Principles can be demonstrated.
- While it may be preferable for harmonization purposes to use international standards, it may be appropriate for Regulatory Authorities to accept the use of national/regional standards or industry standards as a means of demonstrating compliance.
- Standards Bodies developing or revising standards for use with medical devices should consider the suitability of such standards for demonstrating compliance with the Essential Principles and to identify which of the Essential Principles they satisfy.
- The use of standards should preferably reflect current, broadly applicable technology while not discouraging the use of new technologies.
- Standards may represent the current state of art in a technological field. However, not all devices, or elements of device safety and/or performance may be addressed by recognized standards, especially for new types of devices and emerging technologies.

---

[1] As developed by the Global Harmonization Task Force Study Group 1.

'Recognized' Standards[2]

Regulatory Authorities should have or develop a procedure for 'recognition' of voluntary standards and public notification of such recognition. The process of recognition may vary from country to country. An organization may be created/designated by a country or community to develop recognized voluntary standards (e.g. CEN in Europe), or recognition may occur by publication of existing voluntary standards that a Regulatory Authority has found will meet specific premarket requirements (e.g. USA).

Persons intending to market medical devices should obtain information from the relevant Regulatory Authority, Conformity Assessment Body or other authorized third party, or through official publications, on any standards recognized by the Regulatory Authority.[3]

The term 'Recognized Standard' does not imply that such a standard is mandatory.

Compliance with recognized standards may be used (if the manufacturer chooses) to demonstrate compliance with the relevant Essential Principles for Safety and Performance of Medical Devices and/or specific premarket requirements and/or other requirements of the Regulatory Authority.

Alternatives to Recognized Standards

The use of standards is voluntary, except in those particular cases where certain standards have been deemed mandatory by the Regulatory Authority. Manufacturers should be free to select alternative solutions to demonstrate their medical device meets the relevant Essential Principle. Manufacturers may use 'non-recognized' standards, in whole or in part, or other methods. Alternative means of demonstrating compliance with the Essential Principles may include, for example:

- national and international standards that have not been given the status of a 'recognized standard' by the Regulatory Authority
- industry standards
- internal manufacturer standard operating procedures developed by an individual manufacturer and not related to international standards
- current state of the art techniques related to performance, material, design, methods, processes or practices.

The acceptability of such other solutions should be demonstrated.

---

[2] In European Directives 'recognized' standards are known as 'harmonized' standards.

[3] International standards that may be useful in demonstrating compliance with the Essential Principles may be found in overview documents of international standards organizations e.g. ISO TR16142.

Technical documentation

The manufacturer should retain or be able to provide documentation to demonstrate that the device complies with the selected standard or alternative means of meeting the Essential Principles.

Documentation may include for example, the standard itself (or the alternative means used), how it was applied, deviations, test results and/or other outputs.

When a standard is not applied, or is not applied in full, the manufacturer should retain, and submit where appropriate, data or information to demonstrate:

- that compliance with the Essential Principles has been achieved by other means, and if applicable,
- that the parts of the standard that were not applied were not pertinent to the particular device in question.

A declaration of conformity to a recognized standard may be documented in a Summary Technical File,[4] and submitted where appropriate, in lieu of the technical documentation. The format of the declaration of conformity may vary from country to country but it is desirable that a common format be developed.

General information on use of standards

General considerations

The following considerations should be kept in mind when developing a regulatory programme using voluntary standards:

- Standards represent the opinion of experts from industry, regulators, users and other interested parties.
- Standards are based on current scientific knowledge and experience.
- Innovation may present unanticipated challenges to experience.
- Rigid and mandatory application of standards may deter innovation.
- Operation of a quality system, subject to assessment, has become widely acknowledged as a fundamental and effective tool for the protection of public health.
- Quality systems include provisions that address both innovation and experience.
- Such provisions include field experience, risk analysis and management, phased reviews, documentation and record keeping as well as the use of product and process standards.

[4]Refer to the GHTF document: Summary Technical File for the Pre-Market Documentation of Conformity with Requirements for Medical devices (GHTF.SG1.N011R9).

Types of standards

Standards are created and published by national or international standards organizations or by Regulatory Authorities. Examples of international standards bodies are IEC and ISO, of regional bodies are CEN and CENELEC, and of national bodies are Deutsches Institut für Normung, the British Standards Institute, the American National Standards Institute, ASTM, AAMI, the Japanese Industrial Standards Committee and European and national Pharmacopoeias.

Standards are produced for different reasons and are used in some countries as regulatory requirements rather than being voluntary.

Various terms are used to describe the characteristics of a standard. These are not necessarily mutually exclusive. For example:

– **basic safety standards** (also known as horizontal standards)—standards indicating fundamental concepts, principles and requirements with regard to general safety aspects applicable to all kinds or a wide range of products and/or processes (e.g., standards concerning risk assessment and control of medical devices)
– **group safety standards** (also known as semi-horizontal standards)— standards indicating aspects applicable to families of similar products and/or processes making reference as far as possible to basic safety standards (e.g. standards concerning sterile or electrically-powered medical devices), and
– **product safety standards** (also known as vertical standards)—standards indicating necessary safety aspects of specific products and/or processes, making reference, as far as possible, to basic safety standards and group safety standards (e.g., standards for infusion pumps or for anaesthetic machines).

Appendix 1. Sources of standards

1. Sources of standards

International standards may be obtained from national standard bodies and possibly at a financial cost. Information is available also from the ISO and IEC Internet web sites.

IEC: International Electromedical Commission

Address: Central Office of the IEC
             3, rue de Varembe
             P.O. Box 131
             CH-1211 Geneva 20
             Switzerland

Telephone: (+41) 22 919 02 11

Fax: (+41) 22 919 03 00

Web Site: http://www.iec.ch

ISO: International Organization for Standardization

Address: 1, rue de Varembe
              Case postale 56
              CH-1211 Geneve 20
              Switzerland

Telephone: (+41) 22 749 01 11

Fax: (+41) 22 733 34 30

e-mail: central@iso.ch

Web Site: http://www.iso.ch

CEN: CEN Central Secretariat

Address: rue de Stassart, 36
              1050 Brussels
              Belgium

Telephone: (+32) 2 519 68 11

Fax: (+32) 2 519 68 19

Web Site: http://www.stri.is/cen

CENELEC: Comité Européene de Normalisation Electrotechnique

Address: Comité Européene de Normalisation Electrotechnique
              Rue de Stassart, 35
              B-1050 Brussels
              Belgium

Telephone: (+32) 2 519 68 71

Fax: (+32) 2 519 69 19

Web Site: http://server.cenelor.be

Others: see links to above organizations for other national or regional standardization organizations.

Appendix 4

Adverse event reporting guidance for the medical device manufacturer or its authorized representative

Endorsed by The Global Harmonization Task Force

29 June 1999

The document herein was produced by the Global Harmonization Task Force, a voluntary group of representatives from medical device regulatory agencies and the regulated industry. The document is intended to provide non-binding guidance to regulatory authorities for use in the regulation of medical devices, and has been subject to consultation throughout its development.

There are no restrictions on the reproduction, distribution or use of this document; however, incorporation of this document, in part or in whole, into any other document, or its translation into languages other than English, does not convey or represent an endorsement of any kind by the Global Harmonization Task Force.

The Global Harmonization Task Force: GHTF/FD: 99-7

GLOBAL HARMONIZATION TASK FORCE STUDY GROUP 2

ADVERSE EVENT REPORTING GUIDANCE FOR THE MEDICAL DEVICE MANUFACTURER OR ITS AUTHORIZED REPRESENTATIVE

Introduction

The objective of the adverse event reporting and subsequent evaluations is to improve protection of the health and safety of patients, users and others by disseminating information which may reduce the likelihood of, or prevent repetition of adverse events, or alleviate consequences of such repetition.

This document has been created by the Global Harmonization Task Force Study Group 2: Medical Device Vigilance/Post Market Surveillance. Study Group 2 is made up of representatives of the regulatory authorities of the USA, Europe, Canada, Japan and Australia as well as European, US, Canadian and Japanese industry representatives.

For the purpose of this document, the term 'manufacturer' must be understood as including the manufacturer, its authorized representative or any other person who is responsible for placing the device on the market.

The existing regulatory requirements of the participating countries involved in SG2 require medical device manufacturers to notify NCAs of certain adverse events.

This document represents a global model, which provides guidance on the type of adverse events associated with medical devices that should be reported by manufacturers to a National Competent Authority (NCA). It has been elaborated on the basis of the regulatory requirements existing in the participating member countries.

The information and guidance contained herein represents a model, which may not reflect current regulatory requirements. Even if the present reporting criteria of the participating countries are very similar, they are not identical. This document provides a future model towards which those existing systems should converge. The principles laid down in this document should be considered in the development or amendment of regulatory systems in the participating countries or other countries.

In order to improve the monitoring of the performance of medical devices placed on their market, NCA should encourage the reporting of adverse events by the users. Such reports may be addressed either directly to the NCA, or to the manufacturer, or to both depending on national practices. Where the user informs directly the NCA about an event the NCA should adopt administrative measures to ensure that the pertinent manufacturer is informed without delay of such a notification.

NCAs may require certain adverse events to be reported as soon as possible for public health reasons. In such cases, the report may not contain complete information and should be followed up with a complete report.

The act of reporting an event to a NCA is not to be construed as an admission of manufacturer, user, or patient liability for the event and its consequences. Submission of an adverse event report does not, in itself, represent a conclusion by the manufacturer that the content of this report is complete or confirmed, that the device(s) listed failed in any manner. It is also not a conclusion that the device caused or contributed to the adverse event. It is recommended that reports carry a disclaimer to this effect.

It is possible that the manufacturer will not have enough information to decide definitely on the reportability of an event. In such a case, the manufacturer should make reasonable efforts to obtain additional information to decide upon reportability. Where appropriate, the manufacturer should consult with the medical practitioner or the health-care professional involved, and do his utmost to retrieve the concerned device.

NCAs for medical devices usually have, or are part of larger organizations, which have, a responsibility for the protection of public health. To perform their duties they need information, on a voluntary basis if no regulatory requirement exists, about events coming to the notice of manufacturers which might have relevance to the public health but which are not harmonized under the guidelines set out in this document, e.g. use error. Such notifications might fall outside the national adverse event reporting system and could be made in confidence and with suitable anonymity, depending on respective provisions by the responsible NCA.

As a general principle, there should be a pre-disposition to report rather than not to report in case of doubt on the reportability of an event.

1. Decision process

Any event which meets the three basic reporting criteria listed in sections 1.1 through 1.3 below is considered as an adverse event and should be reported to the relevant NCA.

Reporting may be exempted if any one of the exclusion rules listed in section 2 below is applicable.

However those adverse events involving particular issues of public health concern as determined by the relevant NCA should be reported regardless of exemption criteria (see 1.1.d).

Similarly those adverse events which are subject to an exemption become reportable to the NCA if a change in trend (usually an increase in frequency) or pattern is identified.

Specific rules apply to events involving use error and can be found in section 3.

1.1. An event has occurred

The manufacturer becomes aware of information regarding an event which has occurred with its device.

This also includes situations where testing performed on the device, examination of the information supplied with the device or any scientific information indicates some factor that could lead or has led to an event.

Typical events are:

(a) A malfunction or deterioration in the characteristics or performance.
A malfunction or deterioration should be understood as a failure of a device to perform in accordance with its intended purpose when used in accordance with the manufacturer's instructions.

The intended purpose means the use for which the device is intended according to the data supplied by the manufacturer on the labelling, in the instructions and/or in promotional materials.

(b) An inadequate design or manufacture.
This would include cases where the design or manufacturing of a device is found deficient.

(c) An inaccuracy in the labelling, instructions for use and/or promotional materials.
Inaccuracies include omissions and deficiencies.

Omissions do not include the absence of information that should generally be known by the intended users.

(d) A significant public health concern.
This can include an event that is of significant and unexpected nature such that it becomes alarming as a potential public health hazard, e.g. human immunodeficiency virus (HIV) or Creutzfeldt–Jacob Disease (CJD).

These concerns may be identified by either the NCA or the manufacturer.

(e) Other information becoming available.
This can include results of testing performed by the manufacturer on its products, or by the user prior to being used on the patient, or by other parties.

This can also include information from the literature or other scientific documentation.

1.2. The manufacturer's device is associated with the event

In assessing the link between the device and the event, the manufacturer should take into account:

- The opinion, based on available information, from a healthcare professional;
- Information concerning previous, similar events;
- Other information held by the manufacturer.

This judgement may be difficult when there are multiple devices and drugs involved. In complex situations, it should be assumed that the device was associated with the event.

1.3. The event led to one of the following outcomes

1.3.1. Death of a patient, user or other person

1.3.2. Serious injury of a patient, user or other person

Serious injury (also known as serious deterioration in state of health) is either:

- Life threatening illness or injury.
- Permanent impairment of a body function or permanent damage to a body structure.
- A condition necessitating medical or surgical intervention to prevent permanent impairment of a body function or permanent damage to a body structure.

The interpretation of the term 'serious' is not easy, and should be made in consultation with a medical practitioner when appropriate.

The term 'permanent' means irreversible impairment or damage to a body structure or function, excluding minor impairment or damage.

Medical intervention is not in itself a serious injury. It is the reason that motivated the medical intervention that should be used to assess the reportability of an event.

1.3.3. No death or serious injury occurred but the event might lead to death or serious injury of a patient, user or other person if the event recurs

Some jurisdictions refer to these events as near incidents.

All events do not lead to a death or serious injury. The non-occurrence of such a result might have been due to circumstances or to the timely intervention of health care personnel.

The event is considered 'adverse' if, in the case of reoccurrence, it could lead to death or serious injury.

This applies also if the examination of the device or a deficiency in the information supplied with the device, or any information associated with the device, indicates some factor which could lead to an event involving death or serious injury.

Examples of Reportable Adverse Events

- Loss of sensing after a pacemaker has reached end of life. Elective replacement indicator did not show up in due time, although it should have according to device specification.
- On an X-ray vascular system during patient examination, the C arm had uncontrolled motion. The patient was hit by the image intensifier and his nose was broken. The system was installed, maintained, and used according to manufacturer's instructions.
- It was reported that a monitor suspension system fell from the ceiling when the bolts holding the swivel joint broke off. Nobody was injured in the surgical theatre at that time but a report is necessary (near incident). The system was installed, maintained, and used according to manufacturer's instructions.
- Sterile single use device packaging is labelled with the caution '*do not use if package is opened or damaged*'. The label is placed by incorrect design on inner packaging. Outer package is removed but device is not used during procedure. Device is stored with inner packaging only which does not offer a sufficient sterile barrier.
- A batch of out-of-specification blood glucose test strips is released by manufacturer. Patient uses strips according to instructions, but readings provide incorrect values leading to incorrect insulin dosage, resulting in hypoglycaemic shock and hospitalization.
- Premature revision of an orthopaedic implant due to loosening. No cause yet determined.
- An infusion pump stops, due to a malfunction, but fails to give an alarm. Patient receives under-infusion of needed fluids and requires extra days in hospital to correct.
- Manufacturer of a pacemaker released on the market identified a software bug. Initial risk assessment determined risk of serious injury as remote. Subsequent failure results in new risk assessment by manufacturer and the determination that the likelihood of occurrence of a serious injury is not remote.
- Patients undergoing endometrial ablation of the uterus suffered burns to adjacent organs. Burns of adjacent organs due to thin uterine walls were an unanticipated side effect of ablation.
- Manufacturer does not change ablation device label and fails to warn of this side effect which may be produced when the device is working within specification.
- Healthcare professional reported that during implant of a heart valve, the sewing cuff is discovered to be defective. The valve was abandoned and a new valve was implanted and pumping time during surgery was extended.
- During the use of an external defibrillator on a patient, the defibrillator

failed to deliver the programmed level of energy due to malfunction. Patient died.

- An intravenous set separates, the comatose patient's blood leaks on to the floor, the patient bleeds to death.
- Unprotected ECG cable plugged into the main electricity supply–patient died.
- Fatigue testing performed on a commercialized heart valve bioprosthesis demonstrates premature failure, which resulted in risk to public health.
- After delivery of an orthopaedic implant, errors were discovered in heat treatment records leading to non-conforming material properties, which resulted in risk to public health.
- Testing of retained samples identified inadequate manufacturing process, which may lead to detachment of tip electrode of a pacemaker lead, which resulted in risk to public health.
- Manufacturer provides insufficient details on cleaning methods for reusable surgical instruments used in brain surgery, despite obvious risk of transmission of CJD.

2. Exemption rules

Whenever any one of the following exemption rules is met, the adverse event does not need to be reported to NCA by the manufacturer.

2.1. Deficiency of a new device found by the user immediately prior to its use

Regardless of the existence of provisions in the instruction for use provided by the manufacturer, deficiencies of devices that would normally be detected by the user and where no serious injury has occurred, do not need to be reported.

Examples of non-reportable adverse events

- User performs an inflation test prior to inserting the balloon catheter in the patient as required in the instructions for use accompanying the device. Malfunction on inflation is identified. Another balloon is used. Patient is not injured.
- Sterile single use device packaging is labelled with the caution '*do not use if package is opened or damaged*'. Open package seals are discovered prior to use, device is not used.
- Intravenous administration set tip protector has fallen off the set during distribution resulting in a non-sterile fluid pathway. The intravenous administration set was not used.

2.2. Adverse event caused by patient conditions

When the manufacturer has information that the root cause of the adverse event is due to patient condition, the event does not need to be reported. These conditions could be pre-existing or occurring during device use.

To justify no report, the manufacturer should have information available to conclude that the device performed as intended and did not cause or contribute to death or serious injury. A person qualified to make a medical judgement would accept the same conclusion.

Examples of non-reportable adverse events

- Orthopedic surgeon implants a hip joint and warns against sports-related use. Patient chooses to go water skiing and subsequently requires premature revision due to not following directions.
- Early revision of an orthopaedic implant due to loosening caused by the patient developing osteoporosis.
- A patient died after dialysis treatment. The patient had end-stage-renal disease and died of renal failure.

2.3. Service life of the medical device

When the only cause for the adverse event was that the device exceeded its service life as specified by the manufacturer and the failure mode is not unusual, the adverse event does not need to be reported.

The service life must be specified by the device manufacturer and included in the master record [technical file] or, where appropriate, the instructions for use (IFU). Service life is defined as: the time or usage that a device is intended to remain functional after it is manufactured, placed into use, and maintained as specified. Reporting assessment shall be based on the information in the master record or in IFU.

Reporting of adverse events related to the reuse of devices labelled for single use (or labelled 'for single use only') is handled under Section 3: Use Error.

Examples of non-reportable adverse events

- Loss of sensing after a pacemaker has reached end of life. Elective replacement indicator has shown up in due time according to device specification. Surgical explantation of pacemaker required.
- A drill bit was used beyond end of specified life. It fractured during invasive operation. Operation time was prolonged due to the difficulty to retrieve the broken parts.

2.4. Protection against a fault functioned correctly

Adverse events which did not lead to serious injury or death, because a design feature protected against a fault becoming a hazard (in accordance with relevant standards or documented design inputs), do not need to be reported.

Examples of non-reportable adverse events

- An infusion pump stops, due to a malfunction, but gives an appropriate alarm (e.g. in compliance with relevant standards) and there was no injury to the patient.
- Microprocessor-controlled radiant warmers malfunction and provide an audible appropriate alarm. (e.g. in compliance with relevant standards) and there was no injury to the patient.
- During radiation treatment, the automatic exposure control is engaged. Treatment stops. Although patient receives less than optimal dose, patient is not exposed to excess radiation.

2.5. Remote likelihood of occurrence of death or serious injury

Adverse events which could lead, but have not yet led, to death or serious injury, but have a very remote likelihood of causing death or serious injury, and which have been established and documented as acceptable after thorough risk assessment does not need to be reported.

If an adverse event resulting in death or serious injury occurs, the adverse event is reportable and a reassessment of the risk is necessary. If reassessment determines risk remains remote, previous reports of near incidents of the same type do not need to be reported retrospectively. Decisions not to report subsequent failures of the same type must be documented. Note that change in trend of these non-serious outcomes must be reported as specified in Section 1.

Examples of non-reportable adverse events

- Manufacturer of pacemaker released on the market identified a software bug and determined that the likelihood of occurrence of a serious injury with a particular setting is one out of fifteen million remote. No patients experienced adverse health effects.
- Manufacturer of blood donor sets obtains repeated complaints of minor leaks of blood from these sets. No patient injury from blood loss or infections of staff have been reported. Chance of infection or blood loss has been re-evaluated by manufacturer and deemed remote.

2.6. Expected and foreseeable side effects

Side effects which are clearly identified in the manufacturer's labelling or are clinically well known as being foreseeable and having a certain functional or numerical predictability when the device was used as intended need not be reported.

Some of these events are well known in the medical, scientific, or technology field; others may have been clearly identified during clinical investigation and labelled by the manufacturer.

Documentation, including risk assessment, for the particular side effect should be available in the device master record prior to the occurrence of adverse events: manufacturer can not conclude in the face of events that they are foreseeable unless there is prior supporting information.

Examples of non-reportable adverse events

- A patient receives a second-degree burn during the use in an emergency of an external defibrillator. Risk assessment documents that such a burn has been accepted in view of potential patient benefit and is warned in the instructions for use. The frequency of burns is occurring within range specified in the device master record.
- A patient has an undesirable tissue reaction (e.g. nickel allergy) previously known and documented in the device master record.
- Patient who has a mechanical heart valve developed endocarditis ten years after implantation and then died.
- Placement of central line catheter results in anxiety reaction and shortness of breath. Both reactions are known and labelled side effects.

2.7. Adverse events described in an advisory notice

Adverse events that occur after the manufacturer has issued an advisory notice need not be reported individually if they are specified in the notice. Advisory notices include removals from the market, corrective actions, and product recalls. The manufacturer should provide a summary report, the content and frequency of which should be agreed with the relevant NCA.

Example of non-reportable adverse events

- Manufacturer issued an advisory notice and recall of a coronary stent that migrated due to inadequate inflation of an attached balloon mechanism. Subsequent examples of stent migration were summarized in quarterly reports concerning the recall action and individual adverse events did not have to be reported.

2.8. Reporting exemptions granted by NCA

Common and well-documented events may be exempted by NCA from report-ing or changed to periodic reporting upon request by the manufacturer.

3. Use error

The reportability of adverse events involving use error is not globally harmo-nized. Reportability is subject to regulatory requirements of the relevant NCA.

The term 'use error' covers unintentional and intentional misuse.

Some jurisdictions may require reporting of adverse events involving use error even though the events did not occur in their own jurisdiction.

NCAs generally prefer to receive use error reports, especially those involving death or serious injury, on a voluntary basis.

The reprocessing and re-use of devices labelled by the manufacturer for single use ('single use only') is a common practice. However, since this is clearly outside the intended use of the device as stated by the manufacturer, reports of adverse events potentially associated with re-use should be treated as reports of use error.

Similarly, adverse events due to the use of any device for clinical situations not intended by the manufacturer (also called 'off label' use) should be treated similar to use error.

Appendix 5

Some useful web sites

Europe

European Commission:
  http://www.europa.eu.int

EC medical devices:
  http://europa.eu.int/comm/dg03/directs/dg3d/d2/meddev/meddev.htm

EC harmonized standards:
  http://europa.eu.int/comm/enterprise/newapproach/standardization/
  harmstds/reflist/meddevic.html

EC Notified Bodies:
  http://www.dimdi.de/engl/mpgengl/fr-mpgdte.htm (non-German)
  http://www.dimdi.de/germ/mpg/bs-liste.htm (German)

New Approach Guide:
  http://europa.eu.int/comm/enterprise/newapproach/legislation.htm

European Organisation for Testing and Certification:
  http://www.eotc.be

EUDAMED Data Base:
  http://www.dimdi.de

CEN:
  http://www.cenorm.be

CENELEC:
  http://www.cenelec.org

EUCOMED:
  http://www.eucomed.be

COCIR:
    http://www.cocir.org

EDMA:
    http://www.edma-ivd.be

United States

FDA:
    http://www.fda.gov

CDRH:
    http://fda.gov/cdrh

FDAMA implementation:
    http://www.fda.gov/po/modact97.html

Guidance documents:
    http://www.fda.gov/OHRMS/DOCKETS/98fr/031400.txt

Tracking guidance:
    http://www.fda.gov/cdrh/modact/tracking.pdf

Design control guidance:
    http://www.fda.gov/cdrh/comp/designgd.html

ANSI:
    http://www.ansi.org

AdvaMed:
    http://www.AdvaMed.org

NEMA:
    http://www.nema.org

Japan

MHW:
    http://www.mhw.go.jp

JAAME:
    http://www.jaame.or.jp

JISC:
    http://www.jisc.org

Australia

TGA:
   http://www.health.gov.au/tga/devices/devices.html

Standards Australia:
   http://www.standards.com.au

Canada

Health Canada:
   http://www.hc-sc.ca/hpb

Therapeutic products programme
   http://www.hc-sc.gc.ca/hpb-dgps/therapeut

Standards Council of Canada
   http://www.scc.ca

International

GHTF:
   http://www.ghtf.org

ISO:
   http://www.iso.ch

ISO/CASCO:
   http:www.iso.ch/infoe/comm/CASCO.html

IEC:
   http://www.iec.ch

WHO:
   http://www.who.int

References

Anderson S Conformity Assessment: A View to the Future in FDA and Worldwide. Presentation at Seventh Annual AAMI/FDA International Standards Conference on Medical Devices. Arlington, 25–26 March 1997.

Anhoury P The development of the French medical devices sector. In: Medical Devices; International Perspectives on Health and Safety. Ed. C W D van Gruting, p 399. Elsevier Science BV, Amsterdam, 1994.

Argentina 94 Resolution 255/94. Ministry of Health and Social Action, 7 April 1994.

Australia 89 The Therapeutic Goods Act 1989.

Australia 96a The Therapeutic Goods Amendment Act 1996.

Australia 96b Australian Therapeutic Device Bulletin, April 1996.

Australia 97 The Therapeutic Goods Amendment Act 1997.

Australia 98 Australian Therapeutic Devices Bulletin, No 36-2/98, September 1998.

Australia 00 Australian Therapeutic Devices Bulletin, No 41, May 2000.

Barlow W K A New Public/Private Partnership in the Interest of Public Health. Food and Drug Law Journal, Vol 54, p 363, 1999.

Beech D Developments in Australia. In: Medical Devices; International Perspectives on Health and Safety. Ed. C W D van Gruting, p 353. Elsevier Science BV, Amsterdam, 1994.

Beech D and Donovan A Therapeutic Devices in Australia. Regulatory Affairs Journal (Devices), Vol 3, No 3, p 191, 1995.

Belgium 60 Arreté royal relative à la fabrication, à la préparation et à la distribution en gros des médicaments et à leur dispensation, 6 June 1960.

Belgium 63 Arreté royal portant réglement général de la protection de la population et des travailleurs contre le danger des radiations ionisantes, 28 February 1963.

Belgium 64 Loi sur les médicaments, 25 March 1964.

Bergman and Klefsjo Quality: from customer needs to customer satisfaction. McGraw-Hill, London, 1994.

Bolvary G A   Registration Procedure for Medical Devices in Hungary. Regulatory Affairs Journal (Devices), Vol 4, No 2, p 98, May 1996.

Borchardt K-D   The ABC of Community Law, Luxembourg, Office for Official Publications of the European Communities, 1994.

Brazil 93   Portaria Conjunta Nr 1, 17 May 1993.

Cable J   The Evolution of Australian and New Zealand Device Law. Presentation at the Sixth Global Medical Devices Conference, Lisbon, 8–10 October 1996.

Canada 78   The Medical Devices Regulations. Chapter 871, Consolidated Regulations of Canada, 1978 (as amended).

Canada 85   The Food and Drugs Act. Chapter F-27, Revised Statutes of Canada, 1985.

Canada 98   Medical Devices Regulations. Canada Gazette, Part II, Vol 132, No 11, 27 May 1998.

CEN 91   Agreement on technical cooperation between ISO and CEN (Vienna Agreement). CEN General Assembly Resolution CEN/AG3/ 1990 and ISO Council Resolution 18/1990.

CEN 93   Clinical investigation of medical devices for human subjects. EN 540. European Committee for Standardization, Brussels, 1993.

CEN 94   Validation and routine control of sterilization by ethylene oxide (EN 550), irradiation (EN 552), moist heat (EN 554). European Committee for Standardization, Brussels, 1994.

CEN 97a   Application of EN ISO 9001 to the manufacture of medical devices. EN 46001. European Committee for Standardization, Brussels, 1997.

CEN 97b   Application of EN ISO 9002 to the manufacture of medical devices. EN 46002. European Committee for Standardization, Brussels, 1997.

CEN 97c   Medical devices—Risk analysis. EN 1441. European Committee for Standardization, Brussels, 1997.

CEN 98   Information supplied by the manufacturer with medical devices. EN 1041. European Committee for Standardization, Brussels, 1998.

CEN 99   Quality systems—Medical devices—Particular requirements for the application of EN ISO 9003. European Committee for Standardization, Brussels, 1999.

CEN/CLC 89   General criteria for certification bodies operating quality system certification. EN 45012. European Committee for Standardization/European Committee for Electrotechnical Standardization, Brussels, 1989.

CEN/CLC 94   Internal Regulations. European Committee for Standardization/European Committee for Electrotechnical Standardization, Brussels, 1994

CENELEC 89   IEC-CENELEC Agreement on exchange of technical information between both organisations. Standing CENELEC Document

CLC(PERM)002, 1989 (revised 1992). European Committee for Electro-technical Standardization, Brussels, 1989.

Chai J Y Medical Device Regulation in the United States and the European Union: A Comparative Study. Food and Drug Law Journal Vol 55, No 1, p 57, 2000.

China 96 Catalogue of Medical Device Products Classification. State Pharmaceutical Administration of China. December 1996.

China 00 Regulations for the Supervision and Administration of Medical Devices.

Chung-Hwei Lin Registration of Medical Devices in Taiwan. Regulatory Affairs Journal (Devices), Vol 6, No 3, p 187, August 1998.

Cordonnier J IAPM/EEC relationship, draft AIMD Directive. Private communication, 7 November 1988.

Cutts I Report of Orthopaedic Special Interest Section. Association of British Health-care Industries, London, 14 January 1999.

Czech 66 Law of the Czech National Council No 20/1996 Sb (as amended).

Czech 00 Act for Medical Devices No 123/2000 Sb.

Darday V Presentation at IAPM Meeting on Central European Countries, Brussels, 18 November 1998.

Davis B The Food and Drug Agency Modernization Act of 1997. Proc. Conf. London, 26–29 April 1998, p 28. Regulatory Affairs Professional Society Europe, Brussels.

Deming W E Quality, Productivity and Competitive Position. Massachusetts Institute of Technology, Boston, 1982.

Denmark 82 Staerkstromsreglementet, 1982.

Dick Toni Private communication, 12 March 1998.

Dickinson J G Historic MRA signed with Europe. Medical Device and Diagnostic Industry, p 52, September 1997.

Dodge H F and Romig H G Sampling inspection tables—Single and double sampling. Wiley, New York, 1959.

Doolan P T Report of Commission Working Group on Medical Devices meeting of 15/16 June 1989. EUCOMED, London, 27 June 1989.

Duncan M N The Guide to Good Manufacturing Practice for Sterile Medical Devices and Surgical Products 1981. Journal of Sterile Services Management, February 1986.

EC 65 Council Directive 65/65/EEC on the approximation of provisions laid down by law, regulation or administrative action relating to proprietary medicinal products. Official Journal No L22, p 369, 9 February 1965.

EC 74 Proposal for a Council Directive on the approximation of the laws of the Member States relating to electro-medical equipment used in human or veterinary medicine (COM/74/2178). Official Journal No C33, p 5, 12 February 1975.

EC 76   Council Directive 76/768/EEC on the approximation of the laws of the Member States relating to cosmetic products. Official Journal No L262, p 169, 27 September 1976.

EC 80a   Communication from the Commission to the European Parliament and the Council concerning the consequences of the judgement given by the Court of Justice on February 20 1979 in case 120/78 ('Cassis de Dijon'). Official Journal No C256, p 2, 3 October 1980.

EC 80b   Council Directive 80/836/ EURATOM amending the Directives laying down the basic safety standards for the health protection of the general public and workers against the dangers of ionizing radiation. Official Journal No L246, p 1, 17 September 1980.

EC 83   Council Directive of 28 March 1983 laying down a procedure for the provision of information in the field of technical standards and regulations (83/189/EEC). Official Journal No L109, p 8, 26 April 1983.

EC 84a   Council Directive of 17 September 1984 on the approximation of the laws of the Member States relating to electro-medical equipment used in human or veterinary medicine (84/539/EEC). Official Journal No L300, p 179, 9 November 1984.

EC 84b   Council Directive 84/466/EURATOM laying down basic measures for the radiation protection of persons undergoing medical examination or treatment. Official Journal No L265, p 1, 5 May 1984.

EC 85a   Council Directive 85/374/EEC on the approximation of the laws, regulations and administrative provisions of the Member States concerning liability for defective products. Official Journal No L210, p 29, 7 August 1985.

EC 85b   Council Resolution of 7 May 1985 on a new approach to technical harmonisation and standards (85/C 136/01). Official Journal No C136, p 1, 4 June 1985.

EC 85c   Completing the Internal Market. White Paper from the Commission to the European Council (COM(85) 310 final), 28–29 June 1985.

EC 86   Single European Act. Official Journal No L169, p 1, 29 June 1987.

EC 88   Commission Proposal for a Council Directive on the approximation of the laws of the Member States relating to active implantable medical devices (COM(88) 717). Official Journal No C14, p 4, 18 January 1989.

EC 89a   Council resolution of 21 December 1989 on a global approach to conformity assessment (90/C 10/01). Official Journal No C10, p 1, 16 January 1990.

EC 89b   Council Directive 89/686/EEC on the approximation of the laws of the Member States relating to personal protective equipment. Official Journal No L399, p 18, 30 December 1989.

EC 89c   Council Directive 89/336/EEC on the approximation of the laws of the Member States relating to electromagnetic compatibility. Official Journal No L139, p 19, 23 May 1989 [amended by Directives 92/31/EEC and 93/68/EEC].

EC 89d   Council Directive 89/392/EEC on the approximation of the laws of the Member States relating to machinery. Official Journal No L183, 29 June 1989.

EC 90a   Council Decision of 13 December 1990 concerning the modules for the various phases of the conformity assessment procedures (90/683/ EEC). Official Journal No L380, p 13, 31 December 1990.

EC 90b   Council Directive of 20 June 1990 on the approximation of the laws of the Member States relating to active implantable medical devices (90/ 385/EEC). Official Journal No L189, p 17, 20 July 1990.

EC 90c   Standardization request to CENELEC concerning harmonized standards relating to active implantable medical devices. BC/CENE-LEC/02/89. 19 March 1990.

EC 90d   Joint standardization request to CEN/CENELEC concerning harmonized standards relating to horizontal aspects in the field of medical devices. BC/CEN/CLC/9/89. 19 March 1990.

EC 91a   Proposal for a Council Directive concerning medical devices (COM(91) final). Official Journal No C237, p 3, 12 September 1991.

EC91b   Commission Communication on the development of European Standardization ('the Green Paper'). COM (90) 456 final. Official Journal No C20, p 1, 28 January 1991.

EC 91c   Communication from the Commission to the Council on Standardization in the European Economy. Official Journal No C96, p 2, 15 April 1992.

EC 92a   Revised Proposal for a Council Directive concerning medical devices. Official Journal No C251, p 40, 28 September 1992.

EC 92b   The Treaty on European Union (The Maastricht Treaty). 7 February 1992.

EC 92c   The Agreement on the European Economic Area of 2 May 1992. Official Journal No L1, p 3, 1994.

EC 92d   Council Resolution of 18 June 1992 on the role of European standardization in the European economy. Official Journal No C173, p 1, 9 July 1992.

EC 93a   Council Directive 93/42/EEC of 14 June 1993 concerning medical devices. Official Journal No L169, p 1, 12 July 1993.

EC 93b   Council Decision of 22 July 1993 concerning the modules for the various phases of the conformity assessment procedures and the rules for the affixing and use of the CE conformity marking, which are intended to be used in the technical harmonisation directives (93/465/ EEC). Official Journal No L220, p 23, 30 August 1993.

EC 93c   Council Directive 93/68/EEC amending directives 87/404/EEC (simple pressure vessels), 88/378/EEC (safety of toys), 89/106/EEC (construction products), 89/336/EEC (electromagnetic compatibility), 89/392/EEC (machinery), 89/686/EEC (personal protective equipment), 90/384/EEC (non-automatic weighing instruments), 90/385/EEC (active implantable medical devices), 90/396/EEC (appliances burning gaseous fuels), 91/263/EEC (telecommunications terminal equipment), 92/42/EEC (new hot water boilers fired with liquid or gaseous fuels) and 73/23/EEC (electrical equipment designed for use within certain voltage limits). Official Journal No L220, p 1, 30 August 1993.

EC 94a   Guide to the implementation of Community harmonization directives based on the new approach and the global approach. Office for Official Publications of the European Communities, Luxembourg, 1994.

EC 94b   Guidelines relating to the application of the Council Directive 90/385/EEC on active implantable medical devices and the Council Directive 93/42/EEC on medical devices: Working Document: demarcation with other Directives: Directive 89/336/EEC relating to ElectroMagnetic Compatibility Directive 89/686/EEC relating to Personal Protective Equipment. MEDDEV 2.1/4, European Commission, DG III, Brussels, 1 March 1994.

EC 95a   Commission Proposal for a Directive of the European Parliament and the Council on in vitro diagnostic medical devices (COM (95) 130 final). Official Journal No C172, p 21, 7 July 1995.

EC 95b   CERTIF 95/2 rev 1. The European Quality Assurance Standards EN ISO 9000 and EN 45000 in the Community's New Approach legislation. European Commission DG III/B/3, Brussels, 12 September 1995.

EC 95c   Framework for co-ordination and cooperation between notified bodies, Member States and the European Commission under the Community harmonization directives based on the New Approach and the Global Approach. CERTIF 94/6, 1 June 1995.

EC 95d   Communication from the Commission to the Council and Parliament on the broader use of standardization in Community policy. 11/95 EN. European Commission, Brussels, 1995.

EC 96a   Commission Communication in the framework of the implementation of Council Directive of 14 June 1993 in relation to medical devices (96/C245/02). Official Journal No C245, p 2, 23 August 1996.

EC 96b   Guide to the classification of medical devices. MEDDEV 10/93 rev 5, European Commission, Brussels, March 1996.

EC 96c   Commission Working Paper: Database for regulatory data exchange for the implementation of the EC Directives on medical devices. 111/1996/dataexch/doctrav2, Brussels, February 1996.

EC 97a   Treaty of Amsterdam, 2 October 1997.

EC 97b   Amended proposal for a European Parliament and Council Directive on in vitro diagnostic medical devices. Official Journal No C87, p 9, 18 March 1997.

EC 97c   Accreditation and the Community's Policy in the Field of Conformity Assessment. CERTIF 97/4-EN. European Commission DG III/B/4, Brussels, 7 April 1997.

EC 97d The 1996 Single Market Review—Background information for the report to the Council and the European Parliament (the 'Monti Report'). European Commission, Office for Official Publications of the European Communities, Luxembourg 1997.

EC 98a Directive 98/79/EC of the European Parliament and of the Council of 27 October 1998 on *in vitro* diagnostic medical devices. Official Journal No L331, p 1, 7 December 1998.

EC 98b Directive 98/34/EC of the European Parliament and of the Council of 22 June 1998 laying down a procedure for the provision of information in the field of technical standards and information. Official Journal No L204, p 37, 21 July 1998.

EC 98c Guide to the Implementation of Directives based on New Approach and Global Approach. Doc Certif 98/1, European Commission, Brussels, 12 October 1998.

EC 98d Guidelines relating to the demarcation between Council Directive 90/385/EEC on Active Implantable Medical Devices, Council Directive 93/42/EEC on Medical Devices, and Directive 65/65/EEC relating to Medicinal Products and Related Directives. MEDDEV 2.1/3 Rev 5.1, European Commission, Brussels, March 1998.

EC 98e The EN 45000 Series of Standards and the Conformity Assessment Procedures of the Global Approach (Working Document). CERTIF 97/5 EN Rev 1. European Commission, Brussels, 24 March 1998.

EC 98f The EN 45000 Standards, Accreditation and Notification of Notified Bodies. CERTIF 98/4, European Commission, Brussels, 25 March 1998.

EC 98g Code of Conduct for the functioning of the system of notified bodies. Certif Doc 97/1-EN/-REV 3. European Commission, Brussels, 17 July 1998.

EC 98h Guidelines on a Medical Devices Vigilance System. MEDDEV 2.12/1-rev 3, European Commission, Brussels, March 1998.

EC 98i Council Decision of 22 June 1998 on the conclusion of an Agreement on Mutual Recognition between the European Community and the United States of America. Official Journal No L31, p 1, 4 February 1999.

EC 98j Council Decision of 18 June 1998 on the conclusion of an Agreement on mutual recognition in relation to conformity assessment, certificates and markings between the European Community and Australia. Official Journal No L229, p 1, 17 August 1998.

EC 98k Council Decision of 18 June 1998 on the conclusion of an Agreement on mutual recognition in relation to conformity assessment between the European Community and New Zealand. Official Journal No L229, p 61, 17 August 1998.

EC 98l Council Decision of 20 July 1998 on the conclusion of an Agreement on mutual recognition between the European Community and Canada. Official Journal No L280, p 1, 16 October 1998.

EC 98m Accession Partnerships: Bulgaria; Czech Republic; Estonia; Hungary; Latvia; Lithuania; Poland; Romania; Slovakia; Slovenia. Official Journal No C202, p 1, 29 June 1998.

EC 98n Agreement in the form of an exchange of letters between the European Community and the Republic of Poland regarding a Protocol on a European Conformity Assessment Agreement. Official Journal No L237, p 9, 25 August 1998.

EC 99a Standardization mandate to CEN/CENELEC concerning the development of European standards relating to medical devices. European Commission, DG III, Brussels, 9 September 1999.

EC 99b Commission Communication in the framework of the implementation of Directive 98/79/EC of the European Parliament and of the Council of 27 October 1998 on in vitro diagnostic medical devices. Official Journal, No C288, p 41, 9 October 1999.

EC 99c Commission Communication in the framework of the implementation of Council Directive 90/385/EEC of 20 June 1990 in relation to 'Active Implantable Medical Devices', Council Directive 93/42/EEC of 14 June 1993 in relation to 'Medical Devices'. Official Journal No C228, p 42, 9 October 1999.

EC 99d Commission Communication in the framework of the implementation of Council Directive 93/42/EEC of 14 June 1993 in relation to 'Medical Devices'. Official Journal No C228, p 43, 9 October 1999.

EC 99e Survey on quality systems. European Commission, Brussels, 1998.

EC 00a Designation and Monitoring of Notified Bodies within the Framework of EC Directives on Medical Devices. MEDDEV 2.10/2. European Commission, DG III, Brussels, 22 May 2000.

EC 00b Directive 2000/70/EC of the European Parliament and of the Council of 16 November 2000 amending Council Directive 93/42/EEC as regards medical devices incorporating stable derivatives of human blood or human plasma. Official Journal No L313, p 22, 13 December 2000.

EC 00c Background Note on Mutual Recognition Agreements in the medical devices sector. DG ENTR/D14, European Commission, Brussels, 16 June 2000.

EFOMP 91 Report on the Accident of the Electron Therapy Linear Accelerator Sagittaire of the Radiotherapy Department of the Clinical Hospital of Zaragoza, Spain. European Medical Physics News, No 18, p 5, August 1991.

EUCOMED The EUCOMED Report: Recommendations for a Directive on Medical Devices. EUCOMED, London, 1988.

Fairbairn A.G A Systematic Approach to Ventures. IEE Review, Vol 41, No 5, pp 195–198. Institution of Electrical Engineers, London, 1995.

Finland 86 Sahkotarkastuskeskus Elinspektionscentralen Circular KL 118-86, 1986.

France 70 Loi no 70-1318 du 31 décembre 1970 portant réforme hospitalière.

France 73 Loi Royer du 27 décembre 1973 sur la publicité mensongère.

France 75 Loi no 75-1349 du 31 décembre 1975 sur l'emploi de la langue française.

France 82 Arrêté du 9 décembre 1982 relatif à l'homologation des produits et appareils à usage préventive, diagnostique ou thérapeutique.

France 83 Loi no 93-660 du 21 juillet 1983 relative à la protection des consommateurs.

France 87 Loi no 87-575 du 24 juillet 1987 relative aux établissements d'hospitalisation et à l'équipment sanitaire.

France 88 Loi no 88-1138 du 20 décembre 1988 concernant la protection des personnes qui se prêtent à des recherches biomédicales.

France 90 Décret no 90-899 du 1 octobre 1990 relatif à l'homologation de certains produits et appareils à usage préventif, diagnostique ou thérapeutique utilisés en médecine humaine.

France 93 Guide Général de l'homologation des Matériels Médicaux, March 1993.

France 98 Loi no 98-535 du 1 juillet 1998 relative au renforcement de la veille sanitaire et du contrôle de la sécurité sanitaire des produits destinés à l'homme.

Fryback D G and Thornbury J R The Efficacy of Diagnostic Imaging. Medical Decision Making, Vol 11, No 2, p 88, April–June 1991.

GHTF 94 Guidance on Quality Systems for the Design and Manufacture of Medical Devices, Issue No 7. Global Harmonization Task Force, August 1994. Posted on www.ghtf.org 30 October 2000.

GHTF 99 Summary Technical File for Premarket Documentation of Conformity with Requirements for Medical Devices (GHTF.SG1. NO11R10). Global Harmonization Task Force. Draft of 28 December 1999, available on www.ghtf.org.

GHTF 00a Comparison of the Device Adverse Reporting Systems in USA, Europe, Canada, Australia and Japan. (GHTF. SG2-N6R2). Global Harmonization Task Force, posted 25 October 2000 on www.ghtf.org.

GHTF 00b Labelling for Medical Devices. (GHTF.SG1.NO20R5). Global Harmonization Task Force, 15 March 2000, posted on www.ghtf.org.

George W W Medical Technology and Competitiveness in the World Market: Re-inventing the Environment for Innovation. Food and Drug Law Institute, Washington, 38th Educational Conference, 13 December 1994.

George W W The Progress of Global Harmonization and Mutual Recognition Efforts for Medical Devices. Food and Drug Law Journal, Vol 52, Issue No 3, March 1998.

Germany 68 Gesetz über Technische Arbeitsmittel, 24 June 1968.

Germany 76a Gesetz über den Verkehr mit Arzneimitteln, 24 August 1976.

Germany 76b   Verordnung über den Schutz vor Schäden durch ionisierende Strahlen, 13 October 1976.

Germany 85   Verordnung über die Sicherheit medizinisch-technischer Geräte, 14 January 1985.

Germany 88   Eichordnung, 12 August 1988.

Gleason K L and Speights G E   US product liability considerations affecting the medical device industry. In: International Perspectives on Health and Safety. Ed. C W D van Gruting, p 497. Elsevier Science BV, Amsterdam, 1994.

Globe 86   'Man killed by accident with medical radiation.' The Boston Globe, 20 June 1986.

Grunkemeier G L   Clinical Studies of Prosthetic Heart Valves Using Historical Controls. In: Clinical Evaluation of Medical Devices, Ed. K B Witkin, p 83. Humana Press, Totowa, NJ, 1998.

Hastings K   Regulating Medical Devices: A Comparison of US and Foreign Systems and Advantages of the US System Over Others. Food and Drug Administration, Washington, March 1996.

Helsinki 64   Declaration adopted by the 18th World Medical Assembly in Helsinki, Finland, in 1964, as last amended by the 52nd General Assembly, Scotland, October 2000. World Medical Association Inc., 2000. Available on http://www.wma.net.

Higson G R   The UK Department of Health and Social Security's Scientific and Technical Branch. J. Med. Engng. & Technology, Vol 7, No 3, p 130, May/June 1983.

Higson G R   Developments in the United Kingdom. In: Medical Devices: International Perspectives on Health and Safety. Ed. W D Van Gruting, p 391. Elsevier Science BV, Amsterdam, 1994.

Higson G R   The Challenges and Opportunities of Creating an Integrated Global Market for Medical Devices. Biomedical Instrumentation and Technology, Vol 29, No 5, p 441, Sept/Oct. 1995.

Higson G R   Global Harmonisation Activities. The Regulatory Affairs Journal (Devices), Vol 4, No 1, p 1, February 1996.

HIMA   The 1997 Global Medical Technology Update. Health Industry Manufacturers Association, Washington, 1997

Hilz E   Regulatory Harmonization and Standards Development—Opportunities for Cooperation? 7th Annual AAMI/FDA International Standards Conference, Washington, 25–26 March 1997. Reported in Biomedical Instrumentation and Technology, Vol 31, No 5, p 460, Sept./Oct. 1997.

Hirai T   The Impact on Japan. EUCOMED Conference, 1989.

Hirai T and Nagao K   Developments in Japan. In: Medical Devices; International Perspectives on Health and Safety. Ed. C W D van Gruting, p 361. Elsevier Science BV, Amsterdam, 1994.

Hodges C   Product Liability; European Laws and Practice. Sweet and Maxwell, London, 1993.

Holstein H H and Wilson E C   Developments in Medical Device Regulation. In: Fundamentals of Law and Regulation, Vol II. Ed. D G Adams, R M Cooper and J S Kahan, p 257. Food and Drug Law Institute, Washington, 1997.

Horn A   Private communication, 2 November 1998.

Howard A J   Private communication, 22 April 1998.

Hungary 90a   Decree No 14/1990 (IV.17) SZEM on the aptitude testing and certification of hospital and medico-technical products. Ministry of Social Welfare, 1990.

Hungary 90b   Decree No 13/1990 (IV.17) SZEM on the National Institute for hospital and Medical Engineering. Ministry of Social Welfare, 1990.

Hungary 99   Decree No 47/199 EuM.

Huriet C   Rapport d'Information sur les conditions du renforcement de la veille sanitaire et du contrôle de la sécurité sanitaire des produits destinés à l'homme. Annexe au procès verbal de la séance du Senat, 1996.

Huriet C   The French Law of 20 December 1988. Proc. Conf. London, 26–29 April 1998. Regulatory Affairs Professionals Society Europe, Brussels.

Hutt P B, Merrill R A and Kirschenbaum A M   The Standard of Evidence Required for Premarket Approval Under the Medical Device Amendments of 1976. Food and Drug Law Journal Vol 47, No 6, p 605, 1992.

IAPM   Report of European Commission Working Party meetings of 30 and 31 October 1991. International Association of Prosthesis Manufacturers, Paris, 1991.

ICRP 51   International Commission on Radiological Protection. British Journal of Radiology, Vol 24, p 46, 1951.

ICRP 77   Recommendations of the International Commission on Radiological Protection, ICRP Publication 26. Annals of the ICRP, Vol 1, No 2, 1977.

IEC 77   IEC 601-1, Safety of medical electrical equipment—Part 1: General requirements. International Electrotechnical Commission, Geneva, 1977. (Revised 1988).

IEC 94   IEC Technical Report 479-1: Effects of current on human beings and livestock. Part 1: General aspects. International Electrotechnical Commission, Geneva, 1994.

ISO 93   ISO 9000-2, Quality management and quality assurance standards—Part 2: Generic guidelines for the application of ISO 9001, ISO 9002 and ISO 9003. International Organisation for Standardization, Geneva, 1993.

ISO 94a   ISO 9001, Quality systems—Model for quality assurance in design, development, production, installation and servicing. International Organisation for Standardization, Geneva, 1994.

ISO 94b   ISO 9002, Quality systems—Model for quality assurance in production, installation and servicing. International Organisation for Standardization, Geneva, 1994.

ISO 94c ISO 9003, Quality systems—Model for quality assurance in final inspection and test. International Organisation for Standardization, Geneva, 1994.

ISO 94d ISO 8402, Quality management and quality assurance—vocabulary, 2nd edition. International Organisation for Standardization, Geneva, 1994.

ISO 94e ISO 10993 Parts 1–16, Biological evaluation of medical devices. International Organisation for Standardization, Geneva, 1994.

ISO 96a ISO 13485, Quality Systems—Medical devices—Particular requirements for the application of ISO 9001. International Organisation for Standardization, Geneva, 1996.

ISO 96b ISO 13488, Quality Systems—Medical devices—Particular requirements for the application of ISO 9002. International Organisation for Standardization, Geneva, 1996.

ISO 97a Proposed ISO/IEC policy for the development of quality management and quality assurance sectoral documents. ISO/TC 176 N299, International Organisation for Standardization, Geneva, 30 April 1997.

ISO 99a ISO 14969: Guidance on the application of ISO 13485 and 13488. International Organisation for Standardization, Geneva, 1999.

ISO 99b ISO/CD 16142: Medical devices—Guide to the selection of standards in support of recognised essential principles for safety and performance of medical devices (Technical Report type 3). International Organisation for Standardization, Geneva, 12 January 1999.

ISO 00a ISO in figures. International Organisation for Standardization, Geneva, January 2000.

ISO 00b ISO 14971: Medical Devices—Application of risk management to medical devices. International Organisation for Standardization, Geneva, 2000.

ISO 00c ISO 9000:2000: Quality management systems—Fundamentals and vocabulary. International Organisation for Standardization, Geneva, 2000.

ISO 00d ISO 9001:2000: Quality management systems—Requirements. International Organisation for Standardization, Geneva, 2000.

ISO 00e ISO 9004:2000: Quality management systems—Guidelines for performance improvements.

ISO 00f. ISO 15223: Medical devices—Symbols to be used with medical device labels, labelling and information to be supplied. International Organisation for Standardization, Geneva, 2000.

ISO/IEC 91 Guide 2: General terms and their definitions concerning standardization and related activities. International Organisation for Standardization/International Electrotechnical Commission, Geneva, 1991.

ISO/IEC 96a Guide 62: General requirements for bodies operating assessment and certification/registration of suppliers' quality systems.

International Organisation for Standardization/International Electro-technical Commission, Geneva, 1996.

ISO/IEC 96b Guide 61: General requirements for the assessment and accreditation of bodies carrying out certification/registration of suppliers' quality systems.

ISO/IEC 97 ISO/IEC Directives Part 3. Rules for the structure and drafting of International Standards. Third Edition. International Organisation for Standardization/International Electrotechnical Commission, Geneva, 1997.

ISO/IEC 98 ISO/IEC TR 17010, Accreditation of inspection bodies. International Organisation for Standardization/International Electro-technical Commission, Geneva, November 1998.

ISO/IEC 99a ISO/IEC Guide 63, Guide to the development and inclusion of safety aspects in International Standards for medical devices. International Organisation for Standardization/International Electrotechnical Commission, Geneva, 1999.

ISO/IEC 99b Guide 51, Safety aspects—Guidelines for their inclusion in standards. International Organisation for Standardization/International Electrotechnical Commission, Geneva, 1999.

Ischenko A and Mihaylova V Development of medical devices in Russia. In: Medical Devices; International Perspectives on Health and Safety. Ed. C W D van Gruting, p 481. Elsevier Science BV, Amsterdam, 1994.

Italy 27 Disposizioni varie sulla sanita publica, legge 23 giugno 1927, n. 1070.

Italy 78 Circolare n. 100 del 24 novembre 1978 della Direzione Generale del Servizio Farmaceutico—Presidi medico-chirurgici per uso personale.

Japan 60 Pharmaceutical Affairs Law. Law No 145, 1960.

Japan 61a Enforcement Ordinance of the Pharmaceutical Affairs Law. MHW Ordinance No 11, 1961.

Japan 61b Enforcement Regulations of the Pharmaceutical Affairs Law. MHW Ordinance No 1, 1961.

Japan 87 Quality Assurance Standard for manufacture (Good Manufacturing Practices) of Medical Devices. PAB Notification No 87, 28 January 1987.

Japan 92a Good Clinical Practice for Trials on Medical Devices. PAB Notification No 615, 1 July 1992.

Japan 92b On the Revision of Forms of Adverse Drug Reaction Report. PAB Notification No 143, 26 February 1992.

Japan 93 Good Clinical Practice Manual for Medical Devices. MDD Administrative Communication, 1 March 1993.

Japan 94a Pharmaceutical Affairs Law. Law No 50, 29 June 1994.

Japan 94b Quality Assurance Standard for Medical Devices (QA System Standard for Medical Devices). Yakuhatsu No 1128, 28 December 1994.

Japan 94c Guide to Medical Device Registration in Japan. 5th Edition. Yakuji Nippo, Tokyo, 1994.

Japan 95a   Enforcement Ordinance of the Pharmaceutical Affairs Law. MHW Ordinance No 222, 26 May 1995.

Japan 95b   Enforcement Regulations of the Pharmaceutical Affairs Law. MHW Ordinance No 39, 26 June 1995.

Japan 98   Classification of Medical Device and Operations procedure of the documents affixed to applications for approval for medical devices production or importation. Iyaku-shin No 355, MHW, Tokyo, 31 March 1998.

Kahan J S   The Pitfalls of Linking GMP Inspections to Product Clearances. Medical Device and Diagnostic Industry, May 1992.

Kahan J S   FDA's Revised GMP Regulation: The Road to Global Improvement? Med. Device and Diagnostic Industry, May 1994.

Kahan J S   Medical Devices: Obtaining FDA Market Clearance. Parexel International Corporation, Waltham, MA, 1995.

Katkiewicz W   Developments in Poland. In: Medical Devices; International Perspectives on Health and Safety. Ed. C W D van Gruting, p 477. Elsevier Science BV, Amsterdam, 1994.

Kennedy E M   Prescription for Disaster. The Washington Post. p A21, 22 April 1996.

Kent A   Food and Drug Law Institute Seminar, Washington, June 1996. Reported in Clinica, 713, p 9, 8 July 1996.

Kent A   Implementation of the Medical Devices Directives: The UK Regulator's View. Proc. Conf. London 26–29 April 1998. Regulatory Affairs Professionals Society Europe, Brussels.

Kielanowska J   Registration Requirements; Legislation and procedures: Poland. Regulatory Affairs Journal (Devices), Vol 4, No 4, p 321, November 1996.

Larkin R J   Latin America: latest regulatory developments. Presentation at the International Conference on Major Changes in Japan, Germany, France, Canada, China and Brazil. Washington, DC, 9 September 1998. Health Industry Manufacturers Association, Washington, DC.

Last J M (Ed.)   A Dictionary of Epidemiology. Oxford University Press, New York, 1988.

Lindblom D, Rodriguez L and Bjork V O   Mechanical failure of Bjork-Shiley heart valves. J. Thorac. Cardiovasc. Surgery. Vol 48 (Suppl), p S33, 1989.

Maehara Y   Private communication, 16 October 1995.

Mambretti C   Regulation of medical devices in Italy. In: Medical Devices; International Perspectives on Health and Safety. Ed. C W D van Gruting, p 435. Elsevier Science BV, Amsterdam, 1994.

Marcus K   Medical Device registration: South Africa. Regulatory Affairs Journal (Devices), Vol 4, No 4, p 323, November 1996.

Marlowe D   FDA Use of Standards. Biomedical Instrumentation and Technology, Vol 31, No 5, p 459, September/October 1997.

Martinec L   Presentation at IAPM Meeting on Central European Countries, Brussels, 18 November 1998.

MER 95   Boas Práticas de Fabricação de Produtos Médicos. MERCOSUR Resolution Nr.4/95, 1995.

Merrill R.A   FDA and Mutual Recognition Agreements. Food and Drug Law Journal, Vol 53, Issue 1, p 133, 1998.

Michalicek J   Presentation at IAPM Meeting on Central European Countries, Brussels, November 1998.

Mircheva J   Recent Development of Medical Devices Legislation in Eastern European Countries. Proc 4th European Conference and Exhibition, London, April 1998, p 150. Regulatory Affairs Professional Society, Brussels.

Moore R   An overview of European Standardization in the health care sector. In: Medical Devices: International Perspectives on Health and Safety. Ed. C W D van Gruting, p 237. Elsevier Science BV, Amsterdam, 1994.

Muller J H   Medical Devices in South Africa. In: Medical Devices: International Perspectives on Health and Safety. Ed. C W D van Gruting, p 375. Elsevier Science BV, Amsterdam, 1994.

Munsey R R   Trends and Events in FDA Regulation of Medical Devices Over the last Fifty Years. Food and Drug Law Journal, Special 50th Anniversary Issue, p 163, 1995.

NB 98a   Guidance on when clinical investigation is needed. NB-MED/2.7/Rec 1. Vd TÜV Essen, June 1996.

NB 98b   Market surveillance; vigilance—Post-Marketing Surveillance (PMS) post market/production. NB-MED/2.12/Rec 1. Vd TÜV Essen, 1998.

NB 99   Clinical investigations, clinical evaluation. NB—MED/2.7/Rec 3. Vd TÜV Essen, 1999.

Naito M   Japanese Regulation in Practice, Towards Global Harmonization. Presented at the Sixth Global Medical Devices Conference, Lisbon, 8–10 October 1996.

Nobel J J   Adverse effects reporting. In: Medical Devices; International Perspectives on Health and Safety. Ed. C W D van Gruting, p 275. Elsevier Science BV, Amsterdam, 1994.

Nordenberg T   FDA and Medical Devices (an interview with D Bruce Burlington). FDA Consumer, 6 December 1996.

Ohashi J   Marketing Medical Devices in Japan. Medical Device Technology, Vol 9, No 4, p 32, Jan/Feb. 1998.

Pemble L   China Market Update 1998: the Changing Regulatory Environment. Presentation at the International Conference on Major Changes in Japan, Germany, France, Canada, China and Brazil, Washington, DC, 9 September 1998. Health Industry Manufacturers Association, Washington.

Pilot L R and Waldmann D R   Food and Drug Administration Modernization Act of 1997: Medical Device Provisions. Food and Drug Law Journal, Vol 53, Issue 2, p 267, 1998.

Pirovano D   Reported in Clinica, No 764, 7 July 1997.

Pitta R P   Doing Business in the MERCOSUL countries. Presentation at the International Conference on Major Changes in Japan, Germany, France, Canada, China and Brazil, Washington, DC, 9 September 1998. Health Industry Manufacturers Association, Washington, DC.

Poland 91a   Act of August 30 1991 on Health Care Institutions. (Dziennik Ustaw No 91, 14 November 1991).

Poland 91b   Pharmaceutical Products, Medical Materials, Pharmacies, Wholesaler Outlets and Pharmacy Inspection Act of October 10 1991. (Dziennik Ustaw No 105, 19 November 1991).

Poland 94   Directive of the Ministry of Health and Social Care of December 15 1993 concerning registration of pharmaceutical products and medical materials. (Dziennik Ustaw No 6, 17 January 1994).

Popova M   Presentation at IAPM Meeting on Central European Countries. Brussels, 18 November 1998.

Ratnatunga C P, Edwards M-B, Dore C J and Taylor K M   Tricuspid Valve Replacement: UKHVR Results. Presentation at 34th Annual Meeting, Soc. Thoracic Surg. 1998.

Rickards A   The European Pacemaker Registration Card. Stimulation, No 3, p 8, June 1985.

Rickards A and Cunningham D   From Quantity to Quality: The Central Cardiac Audit Database Project. (Draft 1999, to be published).

Riley K   Device regulators and industry fear that new international standards will jeopardise quality systems requirements. Clinica, 900, p 3, 20 March 2000.

Russell I and Grimshaw J   Health Technology Assessment: basis of valid guidelines and test of effective implementation? In: Clinical Effectiveness from Guidelines to Cost-Effective Practice. Ed. M Deighton and S Hitch. Earlybird Publications, Brentwood, 1995.

SA 73a   The Hazardous Substances Act (South Africa) 1973.

SA 73b   Regulations Concerning the Control of Electronic Products. South African Government Gazette, Vol 94, No 3834, 4 April 1973.

SA 89   Regulations Relating to the Sale of Group III Hazardous Substances. South African Government Gazette, Vol 286, No 11823, 14 April 1989.

SA 91a   Regulations Relating to the Sale of Group III Hazardous Substances. South African Government Gazette, Vol 312, No 13299, 14 June 1991.

SA91b   The Medicines and Related Substances Control Amendment Act 1991 (Act No 94 of 1991).

Samuel F E Jr   US medical technology overview. In: Medical Devices; International Perspectives on Health and Safety. Ed. C W D van Gruting, p 329. Elsevier Science BV, Amsterdam, 1994.

Sanderson M   The ISO 9000 Series of Standards: Where do they fit? In: Proc. 1st Int'l Conf. on ISO 9000 and Total Quality Management, 10–12 April 1996. Ed. S K Ho, p 14, De Montfort Univ. Press, Leicester, 1996.

Seddon J   The Case Against ISO 9000. Oak Tree Press, Dublin, 2000.

Segal S   Regulatory Requirements for Clinical Trials of Medical Devices and Diagnostics. In: Clinical Evaluation of Medical Devices: Principles and Case Studies. Ed. K B Witkin, p 59, Humana Press, Totowa, NJ, 1998.

Shenolikar R and Worroll J   The DHSS scheme for the control of cardiac pacemakers. Health Trends, Vol 14, p 20, 1982.

Shewart W A   Economic Control of Manufactured Product. Van Nostrand, New York, 1933.

Slovac 98   Act of 3 April 1998 on Medicinal Products and Medical Devices. Law No 140/1998 Coll of Laws.

Snyder J W   Silicone Breast Implants—Can emerging medical, legal and scientific concepts be reconciled? J Legal Medicine, Vol 18, No 2, p 133, June 1997.

Spain 78a   Real Decreto 908/1978 de 4 de abril de 1978, sobre control sanitario y homologación de material médico terapéutico y correctivo.

Spain 78b   Orden de 21 de julio de 1978 por la que se regula el registro y control de implantes clinicos, terapéuticos y de corrección.

Spain 84   Consumer Protection Act, 1984.

Spain 86   Ley 14/86 de 25 de abril de 1986 General de Sanidad.

Spilker Bert   Guide to Clinical Trials. Raven Press, New York, 1991.

Stokes Ken   Polyurethane Pacemaker Leads: The Contribution of Clinical Experience to the Elucidation of Failure Modes and Biodegradation Mechanisms. In: Clinical Evaluation of Medical Devices: Principles and Case Studies. Ed. K B Witkin. Humana Press, Totowa, NJ, p 233, 1998.

Suppo M   Quality Management Systems: A Reliable Proof of Conformity. Medical Device Technology, p 32, July/August 1997.

Sweden 75   Act Governing the Control of Industrially Sterilised Single-Use Medical Devices (SFS 1975:187), 7 May 1975.

TABD 95   TransAtlantic Business Dialogue, Overall Conclusions, Sevilla, Spain, 10–11 November 1995.

Taylor K M, Gray S-A, Livingstone S and Brannan J J   The United Kingdom Heart Valve Registry. J. Heart Valve Dis. Vol 1, No 2, November 1992.

Tinkler J   Private communication, 6 November 1998.

Trautman K A   The FDA and Worldwide Quality System Requirements Guidebook for Medical Devices. ASQC Quality Press, Milwaukee, 1997.

Tsukamoto H   Private communication, 25 February 1999.

UK 68a   The Trade Descriptions Act 1968.

UK 68b   The Medicines Act 1968.

UK 74   Reporting of Accidents with and Serious Defects in Medicinal Products and other Medical Supplies and Equipment. DHSS, London, June 1974.

UK 79   BS 5750, Quality Systems, British Standards Institution, London, 1979.

UK 81   Guide to Good Manufacturing Practice for Sterile Medical Devices and Surgical Products. HMSO, London, 1981.

UK 82a   Health Equipment Information No 98. DHSS, London, January 1982.

UK 82b   Memorandum of Understanding between the United Kingdom Government and the British Standards Institution on standards. Signed on 24 November 1982 and published in BS 0: Part 2: 1991. British Standards Institution, London 1991.

UK 83   Reporting Accidents with and Serious Defects in Medicinal Products, Buildings and Plant, Equipment and Other Supplies whether Medical or Non-Medical. HN(83)21. DHSS, London, July 1983.

UK 86   Breast Cancer Screening. Report to the Health Ministers of England, Wales, Scotland and Northern Ireland by a Working Group chaired by Sir Patrick Forrest. HMSO, London, 1986.

UK 87a   The Consumer Protection Act 1987.

UK 87b   Guidance Notes for the Protection of Persons against Ionising Radiations Arising from medical and Dental Use. HMSO, London, 1987.

UK 94a   The Medical Devices (Consequential Amendments—Medicines) Regulations 1994 (SI 1994 No 3119).

UK 94b   The Medical Device Regulations 1994 (SI 1994 No 3119).

UK 95   Safeguarding the Public Health. Medical Devices Agency, London, 1995.

UK 97a   MDA's Proposed Compliance Policy. Consultation letter to Trade Associations. Medical Devices Agency, London, 1997.

UK 97b   The New NHS, Modern, Dependable. HMSO, London, 1997.

UK 98a   Guidance on the Medical Devices Vigilance System for CE marked joint replacement implants. Medical Devices Agency, London, 1998.

UK 98b   *In vitro* Diagnostic Medical Devices. Medical Devices Agency Adverse Incident Centre, London, 1998.

UK 98c   Guidance Document on Medical Device Recalls—Draft No 1. Medical Devices Agency, London, 19 October 1998.

UK 99   1998–99 Annual Report and Accounts. Medical Devices Agency. The Stationery Office, London, August 1999.

USA 38   Federal Food, Drug and Cosmetic Act of 1938. (FDCA, Pub. L. No 75-717, 52 Stat. 1040 (1938), as amended 21 U.S.C. 301 *et seq.* (1938)).

USA 68   The Radiation Control for Health and Safety Act of 1968 (RCHS Act, Pub. L. No 90-602, 82 Stat. 1173 (1968) (codified at 42 U.S.C. 263b-n (1968)).

USA 73   Medical Devices: Hearings Before the Subcommittee on Public Health and Environment of the House Comm. on Interstate and Foreign Commerce, 93rd Congress of the United States, 1st Session, 1973.

USA 76a   The Medical Device Amendments of 1976 (MDA, Pub. L. No 94-295, 90 Stat. 539(1976) (codified at 15 U.S.C. 55, 21 U.S.C. 301, 331, 334, 351, 352, 358, 360, 374, 379, 381 (1988)).

USA 76b   House of Representatives Report No 853, 94th Congress of the United States, 2nd Session, 1976.

USA 78   Good Manufacturing Practice Requirements for the manufacture, packing, storage and installation of medical devices—Final Rule. Federal Register 21 July 1978 (43 FR 31 508).

USA 84   Medical Device Reporting Regulations for Manufacturers (49 FR 36326 at 36348, 14 September 1984).

USA 86   Guidance for Clinical Investigations for a PMA. From: The PMA Approval Manual. HHS Publication FDA 87-4214, CDRH, 1986.

USA 90a   The Safe Medical Devices Act of 1990 (SMDA, Pub. L. No 101-629, 104 Stat. 4511 (1990) (codified at 21 U.S.C. 301 note, 321, 333, 351, 360, 383; 42 U.S.C. 263b-263n)).

USA 90b   Federal Register, 25 April 1990 (55 FR 17502).

USA 90c   Device Recalls: A Study of Quality Problems. HHS Publication FDA 90-4235. Center for Devices and Radiological Health, Rockville, 1990.

USA 92   The Medical Device Amendments of 1992 (Pub. L. No 102-300, 106 Stat. 238 (1992) (codified at 21 U.S.C. 301 note, 321, 331, 334, 346a, 352-353, 356-357, 360c-d, 371-372, 372a, 376, 381; 42 U.S.C. 262)).

USA 93a   *Less than the sum of its parts.* Committee on Oversight and Investigations. H.R. Rep. No 103-N, 103rd Congress, 1st Session 5 (1993).

USA 93b   ISO 9000 Policy Implications for FDA. Office of Policy, US Food and Drug Administration, Washington, January 1993.

USA 93c   Final Report of the Committee for Clinical Review. Center for Devices and Radiological Health, Rockville, 1993.

USA 93d   Medical Device Clinical Study Guidance. Center for Devices and Radiological Health, Rockville, 1993.

USA 93e   Draft Replacement Heart Valve Guidance. Prosthetic Devices Branch, Division of Cardiovascular, Respiratory and Neurological Devices, Center for Devices and Radiological Health, Rockville, 1993.

USA 94   International Harmonization; Draft Policy on Standards; Availability. Federal Register Vol 59, No 227, 28 November 1994.

USA 95a   Medical Devices: Medical Device User Facility and Manufacturer Reporting, Certification and Registration, 21 CFR Parts 803 and 807, Federal Register, Vol 60, No 237, 11 December 1995.

USA 95b Proposals to Improve FDA's Medical Device Program and Points of Controversy About the Medical Devices Program. Center for Devices and Radiological Health, Rockville, March 1995.

USA 96a Medical Device Regulation: Too Early to Assess European System's Value as Model for FDA. GAO/HEHS-96-65. General Accounting Office, Washington, March 1996.

USA 96b Medical Devices; Current Good Manufacturing Practice (CGMP) Final Rule; Quality System Regulation 21 CFR Parts 808, 812 and 820, Federal Register, 7 October 1996 (61 FR 52602).

USA 96c Federal Register, 13 April 1996.

USA 97a The Food and Drug Administration Modernization Act of 1997.

USA 97b Mutual Recognition Agreement, US Working Draft. Article III.4.b. US Dept. of Commerce, Washington, DC, 25 February 1997.

USA 98a FDA Modernization Act of 1997: Guidance for the Recognition and Use of Consensus Standards; Availability. Federal Register, 25 February 1998 (63 FR 9561).

USA 98b Criteria for the accreditation of third parties. Federal Register, 22 May 1998 (63 FR 28388).

USA 98c Mutual Recognition of Pharmaceutical Good Manufacturing Practice Inspection Reports, Medical Device Quality System Audit Reports and Certain Medical Device Product Evaluation Reports Between the United States and the European Community; Final Rule. Federal Register, 6 November 1998 (63 FR 60121).

USA 98d Global Harmonization Task Force: Draft Documents on Adverse Event and Vigilance Reporting of Medical Device Events; Availability. Federal Register, 31 August 1998 (63 FR 46227).

USA 98e Guidance on Criteria and Approaches for Postmarket Surveillance. Center for Devices and Radiological Health, Rockville, 2 November 1998.

USA 99a Medical Devices; Global Harmonization Task Force: Summary Technical File Documents for Premarket Demonstration of Conformity with Requirements for Medical Devices; Recommendations on the Role of Standards in the Assessment of Medical Devices; and Recommendations on Medical Device Classification; Availability. Federal Register, 9 August 1999 (64 FR 43196).

USA 99b Letter from Lilian Gill (FDA) to the European Commission, 6 June 1999.

USA 00a Medical Device Reporting: Manufacturer Reporting, Importer Reporting, User Facility Reporting, Distributor Reporting—Final Rule. Federal Register, 26 January 2000 (65 FR 4112).

USA 00b Proposed Rule: Medical Device Tracking. Federal Register, 25 April 2000 (65 FR 24144).

USA 00c   Draft Guidance for Staff, Industry, and Third Parties; Implementation of Third Party Programs Under the FDA Modernization Act of 1997; Availability. Federal Register, 18 July 2000 (65 FR 44540).

USA 01a   Implementation of Third Party Programs Under the FDA Modernization Act of 1997; Final Guidance for Staff, Industry, and Third Parties. CDRH, Rockville, 2 February 2001.

USA 01b   Devices—Inspections of Medical Device Manufacturers Compliance Program Guidance Manual, CP 7382.845; Availability. Federal Register, 7 February 2001 (66 FR 9347).

Van Vleet J D   Prospective Multicenter Clinical Trials in Orthopedics: *Special Concerns and Challenges*. In: Clinical Evaluation of Medical Devices: Principles and Case Studies, Ed. K B Witkin, p 103. Humana Press, Totowa, NJ, 1998.

WHO 87   Proc. 1st Int. Conf. of Medical Device Regulatory Authorities (1996). World Health Organisation, Pan American Health Organisation, Food and Drug Administration, Washington, DC 1987.

Waisbord E, Normand M, Virefleau R, Vadrot D and Pinget M   La procédure d'homologation française. Infu-Systèmes, vol 16, No 3, p 20, October 1989.

Wilkerson Group   Forces Reshaping the Performance and Contribution of the US Medical Device Industry. HIMA, Washington, 1995.

Witkin K B   (a) Clinical trials in Development and Marketing of Medical Devices. In: Clinical Evaluation of Medical Devices: Principles and Case Studies, Ed. K B Witkin, p 3. Humana Press, Totowa, NJ, 1998.

Witkin K B   (b) Introduction to Case Studies. In: Clinical Evaluation of Medical Devices: Principles and Case Studies, Ed. K B Witkin p 79. Humana Press, Totowa, NJ, 1998.

Witkin K B   (c) How to Design a Postmarketing Surveillance Programme. The Regulatory Affairs Journal (Devices), Vol 6, No 2, p 87, May 1998.

Wolf M   Medical equipment technology in Germany. In: Medical Devices; International Perspectives on Health and Safety. Ed. C W D van Gruting, p 417. Elsevier Science BV, Amsterdam, 1994.

WTO 95   Agreement on Technical Barriers to Trade. MTN/FA II-A1A-6. World Trade Organisation, 1995.

Bibliography

Abbott H Managing Product Recalls. Pitman, London, 1991.

ACARD (Advisory Committee on Applied Research and Development). Medical Equipment. HMSO, 1986.

ACOST (Advisory Committee on Science and Technology). A Report on Medical Research and Health. HMSO, 1993.

Adams D G Cooper R M and Kahan J S (Eds) Fundamentals of Law and Regulations, Vol. II, Food and Drug Law Institute, Washington, 1997.

Appelbe G and Wingfield J Dale and Appelbe's Pharmacy Law and Ethics. Pharmaceutical Press, London, 1993.

Barnes R W Clinical Studies: How to successfully resolve clinical problems to achieve FDA approval. Regulatory Affairs, p 29, Spring 1991.

Bennett A R and Dormer R A FDA and Radiological Health: Protecting the Public from Radiation Hazards. In Fundamentals of Law and Regulation, Vol. II, Eds D G Adams, R M Cooper and J S Kahan. Food and Drug Law Institute, Washington, DC, p 299, 1997.

Borchard K-D European Integration—The origins and growth of the European Union. Office for Official Publications of the European Communities, Luxembourg, 1995.

Cecchini P The European Challenge, 1992: The Benefits of a Single Market. Wildwood House, Aldershot, 1988.

Clinica Trends in Medical Device Regulation. Special Supplement, PJB Publications, Richmond, October 1993.

Clinica Europe's Regulatory Labyrinth. Special Supplement, PJB Publications, Richmond, October 1996.

Cockfield, Lord The European Union—Creating the Single Market. Chancery Law Publishing, Chichester, 1994.

Deighan M and Hitch S (Eds) Clinical Effectiveness from Guidelines to Cost-effective Practice. Earlybrave Publications, Brentwood, 1995.

European Commission Council Directive 92/59/EEC on general product safety. Official Journal No L228, p 24, 1992.

European Commission (Draft) Review of European Standardization Policy. Working Paper of DG III, Brussels, 12 December 1995.

European Commission Green Paper—Public procurement in the European Union: Exploring the way forward. Communication adopted by the Commission on 27 November 1996.

Evers P Medical Device Regulation in Europe. Financial Times Healthcare, London, 1996.

FDA PMA Approval Manual. HHS Publication FDA 87-4214. Center for Devices and Radiological Health, Rockville,1986.

FDA Investigational Device Exemptions Manual. HHS Publication FDA 90-4159, Center for Devices and Radiological Health, Rockville, August 1990.

Gelijns A C Innovation in clinical practice; the dynamics of medical technology development. National Academy Press, Washington, DC, 1991.

Haindl H Safety of Medical Devices and Where it Comes From. Medical Device Technology, Vol 12, p 62, Sept./Oct. 1990.

Higson G R The Medical Devices Directives—A Manufacturer's Handbook. Medical Technology Consultants, Staines, 2nd edition 1996.

Higson G R and Howard A J Medical Device Approvals in Europe Today. A Series of Monographs. BRI International, Arlington, 1990.

HIMA White Paper on FDA Medical Device Reform. Health Industry Manufacturers' Association, Washington, 1995.

Hitchins D K Putting Systems to Work. Wiley, 1992.

Hodges C, Tyler M and Abbott H Product Safety. Sweet and Maxwell, London, 1996.

IEC IEC TR 513: Fundamental aspects of safety standards for medical equipment. International Electrotechnical Commission, Geneva, 1994.

Institute of Medicine Assessing Medical Technologies. National Academic Press, Washington, DC, 1985.

JFMDA Pharmaceutical Affairs Law, Extracts of Parts Affecting Medical Devices (Japan). Edited by The Japan Federation of Medical Devices Associations. Yakuji Nippo Ltd, Tokyo, 1998.

Kahan J S Utilization of clinical data in support of US and EU device clearances. Hogan & Hartson L.L.P., Washington, DC.

Landry P, Marshall R T and Grosskopf M (Eds) The ISO 9000 and 13485 Essentials. Canadian Standards Association, Etobicoke, 2nd edition 1998.

Leener H, Prims A and Pinet J Trends in Health Legislation. World Health Organisation. Masson, Paris, 1986.

Louis J-V The Community Legal Order. Office for Official Publications of the European Communities, Luxembourg, 1995.

Merrills J and Fisher J Pharmacy Law and Practice. Blackwell, Oxford, 1995.

Monti M   The Single Market and Tomorrow's Europe. Office for Official Publications of the European Communities, Luxembourg, 1996.

Muir Gray J A   Evidence Based Health Care. Churchill Livingstone, New York, 1997.

Munsey R R   Trends and Events in FDA Regulation of Medical Devices over the Last Fifty Years. Food and Drug Law Journal, 50th Anniversary Edition, p 163, 1995.

Nicholas F   Common Standards for Enterprises. Office for Official Publications of the European Communities, Luxembourg, 1995.

Oliver P   Free Movement of Goods in the EEC. European Law Centre, London, 1983.

Prechal S   Directives in European Community Law. Clarendon Press, Oxford, 1995.

Roscam Abbing H D C   Medical Devices and Safety in Health Care. Health Policy, Vol 8, p 59, 1987.

Roscam Abbing H D C   Medical Device Safety in the European Community—A Legal Perspective. Institute of Health Law. State University of Maastricht, 1991.

Schwamm H   World Trade Needs Worldwide Standards. ISO Bulletin, p 12, September 1997.

Stephens R N   Medical Device Vigilance/Monitoring: European Device Directives Compliance. Interpharm Press, Buffalo Grove, 1997.

Thompson B M and Wang X   Medical Devices in China. The Regulatory Affairs Journal (Devices), Vol 6, No 1, p 25, February 1998.

Walker, J S   Permissible Dose: A History of Radiation Protection in the Twentieth Century. University of California Press, 2000.

Weatherill S and Beaumont P   EC Law, Penguin, Second Edition 1995.

Index

263